Crystallization from solutions and melts

J. A. Palermo and M. A. Larson, editors

G. R. Atwood	H. J. Jensen	S. A. Ring
G. W. Becker, Jr.	Stanley Katz	M. B. Sherwin
Richard C. Bennett	M. A. Larson	Reuel Shinnar
George Burnet	L. M. Litz	G. J. Sloan
Howard M. Dess	R. A. Mercuri	Denis G. Stefango
Vladimiro Ern	M. J. Murtha	Paul D. Stone
Hugh M. Hulburt	A. D. Randolph	Maarten Van Buren

Number 95 1969 Volume 65

Springer Science+Business Media, LLC

Springer Science+Business Media New York
Originally published by American Institute of Chemical Engineers in 1969.
ISBN 978-1-4899-4817-5 ISBN 978-1-4899-4815-1 (eBook)
DOI 10.1007/978-1-4899-4815-1

Library of Congress Card Number 70-90832

TABLE OF CONTENTS

FOREWORD

This collection of papers was developed primarily from a three-session symposium on crystallization held at the Second Joint AIChE-IIQPR meeting, held in Tampa, Florida, May 19–22, 1968. The possibility of such a volume is evidence of the recent increase in interest in analysis of crystallization processes. As some of these papers show, the recent advances in understanding of the crystallization process promise to eventually raise the design and operation of the crystallization process from an almost pure art to a state where more meaningful analyses and calculations can be made.

The first paper in Part I is the text material of a workshop dealing with recent developments in analysis of factors influencing size distribution in continuous crystallizers. This paper lays the theoretical groundwork for the remaining papers in Part I which include theoretical analyses, experimental results, and applications. The first six were given at the Tampa meeting. The remaining two were included to give a more complete picture of the status of this very important aspect of crystallization. The reader should take note of the notation differences among these papers. In the first five, population density is represented by n, growth rate by r, and crystal size by L. In other three, population density is represented by f, growth rate by G, and crystal size by r.

The papers in Part II generally give the state of the art in several aspects of crystallization from the melt. In the main they describe techniques for growing single crystals, purification by crystallization, and multistage crystallization from the melt. The first four papers were given at the Tampa meeting. The fifth paper was given at the AIChE meeting in Mexico City, September, 1967.

The efforts of Dr. A. D. Randolph, Dr. John Stevens, and Dr. Joseph Estrin in organizing and conducting the three sessions at Tampa were greatly appreciated. The extra efforts required of Dr. Stevens and Dr. Estrin because of the absence of Joe Palermo and myself contributed considerably to the success of this undertaking.

The senior editor of this volume, Dr. J. A. Palermo, died suddenly during the preparation of this material. He was responsible for the conception of this volume and carried the primary work load. He also was the driving force behind the very successful three-session symposium at Tampa from which most of these papers were drawn. The recent increase in activity of the crystallization group of the National AIChE Program Committee largely occurred during his term as chairman. In addition, Dr. Palermo for several years was responsible for the annual review of crystallization work published in *Industrial and Engineering Chemistry*. The field of crystallization research and the profession of chemical engineering has lost an energetic and purposeful contributor. I personally have suffered the loss of a trusted friend who had repeatedly offered me help and encouragement in my work.

M. A. Larson, *editor*
Iowa State University of Science and Technology

SIZE DISTRIBUTION ANALYSIS IN CONTINUOUS CRYSTALLIZATION

M. A. Larson
and
A. D. Randolph

The basic ideas of the population balance are presented along with the use of the population balance in relating system parameters and crystallization kinetics with size distribution. The principles are illustrated in the context of classification and fines destruction systems. Example problems are provided for the reader.

POPULATION BALANCE

In the analysis of crystallization systems, an important consideration in studying the factors influencing size distribution is the method of characterizing the distribution. In most design procedures currently in use, the size distribution is characterized by the mass of crystals in a given size range. This type of distribution is obtained easily from a screen analysis of the crystals. It is of little use, however, when the influence of nucleation rate on size distribution is under study. As it turns out, the nucleation kinetics is the primary factor limiting crystal size obtainable in any continuous crystallization process. It is, therefore, of considerable interest to have an analytical approach which provides a method for characterizing the crystal distribution and which will, at the same time, facilitate the determination of nucleation effects on size distribution.

In the consideration of any process, the chemical engineer is quick to recognize that material and energy balances must be satisfied. Such balances are therefore made, and they usually result in mathematical expressions suitable for design purposes. Coupled with the appropriate rate equations, they are used to define the system. The classic McCabe-Thiele method for distillation column design is a simple example of the mass balance used for design purposes. In crystallization processes, these balances must also be satisfied, but one additional conservation law is needed if the size distribution is to be fully defined. This balance is called a *numbers* or *population balance.*

Phase change in solution crystallization results not in a single homogeneous mass of solid material, but in a collection of discrete particles of various sizes, shapes, and ages. This particle distribution results from birth (nucleation) of new particles and subsequent growth. This distribution is usually characterized as some function of crystal size rather than shape or age, hence, the term *size distribution.* The size distribution is dependent on the birth rate, the growth rate, and, when applicable, the death rate or the rate of disappearance. It is necessary, therefore, to know what factors influence these rates. It must be remembered that the rates in question are those pertaining to the individual particles and not those pertaining to the gross mass of material. In order to properly study the various effects, a study must be made of the effects on the particles, not the collective mass. It is, therefore, necessary to enumerate the population of particles with a given size, to define the manner in which they are born, and to determine what factors influence their growth rate.

The development of such a characterization can best be attacked by first recognizing that, like mass and energy, the number of discrete particles must also be conserved in any dispersed system, and that given the proper representation of birth and death rates, all particles can be accounted for. Such an accounting is called a *population balance* and just as in the case of other balances, results in a mathematical expression useful in characterizing some aspect of the system. In this case it characterizes the size distribution.

Iowa State University of Science and Technology, Ames, Iowa. A. D. Randolph is with the University of Arizona, Tucson, Arizona.

In crystallizing systems, we are interested primarily in characterizing the population according to size. Suppose there exists a collection of particles in suspension of uniform shape factor, that is to say, the various spacial dimensions of all the particles have a one to one correspondence. For such a collection, one dimension of the particle will adequately characterize the size of the crystal. For the purposes of this development, the dimension can be the screen mesh size through which it just passes. Of course other dimensions could also be used.

POPULATION DENSITY

To facilitate the formulation of the population balance, it is necessary to define a continuous variable to represent the discrete distribution. To do this the density function of the statistician is used. In simple terms, the density function to be used is a representation of the number of crystals in a given size range.

Consider now the number of crystals in a size range from L_1 to L_2 and let $L_2 - L_1$ be represented by ΔL. The number of crystals in this size range is ΔN. The number of crystals is then divided by the width of the size range, and a size range of differential width is considered. Thus

$$\lim_{\Delta L \to 0} \frac{\Delta N}{\Delta L} = \frac{dN}{dL} = n \quad (1)$$

The quantity n thus defined is called the *population density*. Clearly, the value of n will be determined by the crystal size at which the interval dL is taken. Consequently, n is a function of L.

Graphically, n may be determined by differentiating a cumulative number plot as shown in Figure 1.

In order to return to the distribution in terms of numbers, it is only necessary to integrate n over the range of L:

$$\int_{L_1}^{L_2} ndL = \text{number of crystals in with sizes from } L_1 \text{ to } L_2 \quad (2)$$

The population density will now be used in the formulation of a numbers or population balance.

CONTINUOUS MIXED SUSPENSION MIXED PRODUCT REMOVAL CRYSTALLIZER

To begin the study of crystallization systems via the population balance, it is convenient to use the concept of a continuous mixed suspension mixed product removal (CMSMPR) crystallizer. Such a system has the advantage that it can be examined easily under steady state conditions. A further advantage lies in the fact that such crystallizers or closely related crystallizers are in widespread industrial use. Consequently, results from the study of such systems are highly useful. It should be pointed out that it is extremely important to be able to measure and study crystallization under steady state conditions.

Consider now a well-stirred vessel which is continuously fed a solution of constant composition as shown in Figure 2. The vessel is cooled so that crystallization takes place. (Actually the restriction of cooling as a method of creating supersaturation is not in general necessary; however, for ease in understanding, this discussion will be limited to such systems.) A well-mixed slurry with unclassified product is removed continuously. It is further assumed that no attrition occurs, and that no particles enter with the feed. A steady state numbers balance (population balance) for an arbitrary

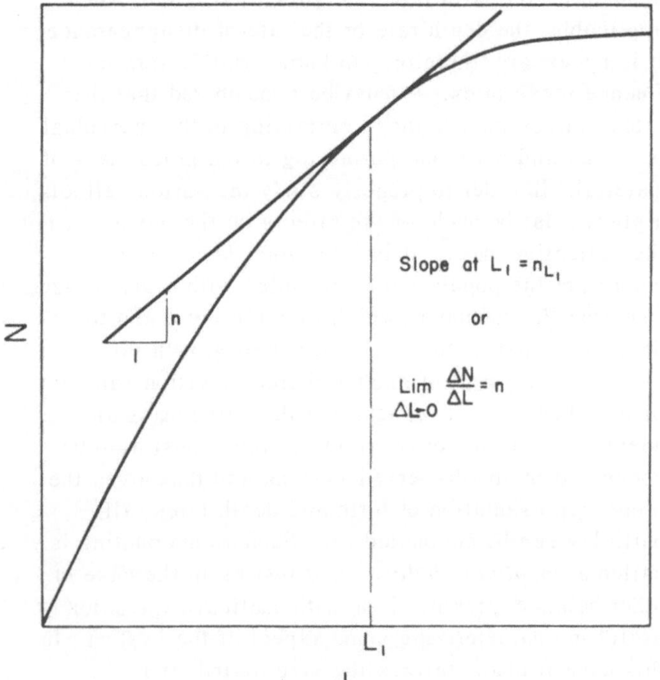

Fig. 1. Cumulative numbers.

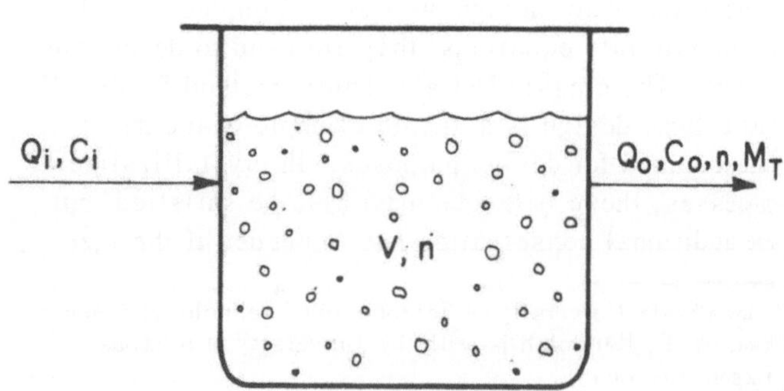

Fig. 2. Continuous crystallizer.

size range L_1 to L_2 for an arbitrary time interval Δt in terms of population density is

Input		Output	
$Vr_1n_1\Delta t$ =	$Vr_2n_2\Delta t$ +	$Q\bar{n}\,\Delta L\,\Delta t$	(3)
number of crystals growing into range over the time interval	number of crystals growing out of the range over the time interval	number of crystals in the size range removed from crystallizer	

The population density n has dimensions of numbers per unit volume of slurry.

Dropping the Δt's and rearranging, we get

$$\frac{V(r_2n_2 - r_1n_1)}{\Delta L} + Q\bar{n} = 0 \tag{4}$$

In the limit as $\Delta L \longrightarrow 0$

$$V\frac{d(rn)}{dL} + Qn = 0 \tag{5}$$

Letting V/Q equal T, the drawdown time, and assuming that McCabe's ΔL law holds, that is, growth is independent of size, we obtain

$$\frac{dn}{dL} + \frac{n}{rT} = 0 \tag{6}$$

If n° is the population density of the zero size particles or nuclei, then integration of (6) gives

$$\int_{n^\circ}^{n}\frac{dn}{n} = \int_0^L \frac{dL}{rT} \tag{7}$$

$$n = n^\circ \exp(-L/rT) \tag{8}$$

Equation (8) gives the functional relationship between L and n and thus characterizes the size distribution.

MOMENTS OF THE DISTRIBUTION

At this point it is interesting to note how other distribution representations can be obtained from the above representation. From the definition of population density n, it was shown by Equation (2) that ndL is the number of crystals in size range dL and that the integral of ndL is the total number of crystals up to size L:

$$N = \int_0^L ndL = \int_0^L n^\circ \exp(-L/rT)\,dL$$
$$= n^\circ rT\,(1 - \exp(-L/rT) \tag{9}$$

It should be noted in passing that Equation (9) is the 0th moment of the distribution n. For L large ($L \longrightarrow \infty$), the above equation reduces to the total number of crystals and becomes

$$N_T = n^\circ rT \tag{10}$$

If the cumulative length is desired, the first moment is formulated by recognizing that $LndL$ represents the total length (if crystals are laid end to end) of all crystals in size range:

$$\mathfrak{L} = \int_0^L LndL = \int_0^L n^\circ L \exp(-L/rT)\,dL$$
$$= n^\circ rT\,(rT\,(1 - \exp(-L/rT)) - L\exp(-L/rT)) \tag{11}$$

As L becomes large, the total length becomes

$$\mathfrak{L}_T = n^\circ (rT)^2 \tag{12}$$

It follows that the second moment is related to the total surface area of the crystals. Here, however, it is necessary to introduce a shape factor k_A for area

$$A = k_A \int_0^L L^2 ndL$$
$$= k_A n^\circ rT\,[2\,(rT)^2\,1 - \exp(-L/rT) - (rTL + L^2)\exp(-L/rT)] \tag{13}$$

or, as L becomes large

$$A_T = 2k_A n^\circ (rT)^3 \tag{14}$$

The cumulative mass distribution can be formulated by using the third moment (volume), a shape factor, for volume, and the crystal density:

$$M = k_v\rho \int_0^L nL^3dL \tag{15}$$

$$= n^\circ k_v\rho\, rT\,[6\,(rT)^3\,[1 - \exp(-L/rT)] - [6\,(rT)^2 L + 3rTL^2 + L^3]\exp(-L/rT)] \tag{16}$$

or in the limit

$$M_T = 6k_v\rho\, n^\circ (rT)^4 \tag{17}$$

For a simpler formulation in dimensionless form, let $x = L/rT$; then

$$\frac{N}{N_T} = 1 - \exp(-x) \tag{18}$$

$$\frac{\mathfrak{L}}{\mathfrak{L}_T} = 1 - (1 + x)\exp(-x) \tag{19}$$

$$\frac{A}{A_T} = 1 - (1 + x/2 + x^2/2)\exp(-x) \tag{20}$$

$$\frac{M}{M_T} = 1 - (1 + x + x^2/2 + x^3/6)\exp(-x) \tag{21}$$

The usefulness of these distributions will be discussed later.

NUCLEATION AND GROWTH RATES

The actual crystal sizes obtained in a crystallizer operated under the constraints indicated are dependent on the parameters in Equation (8). The drawdown time T is arbitrary, but the nuclei population density and the growth rate are related to the fundamental crystallization kinetics. The concept of the CMSMPR crystallizer will now be used to show how the kinetics affect the size distribution and how power law kinetic models can be obtained from experimental data.

Consider such a crystallizer operating at steady state. The size distribution of the product crystals should fit Equation (8). If this is so, then the kinetics can be determined by the following procedures.

From a screen analysis of the product crystals, the population density at the mean size of the various size fractions can be calculated by the following equation:

$$n = \frac{w}{\rho k_v \overline{L}^3 \Delta L} \tag{22}$$

ΔL is the difference between the size of the screen under consideration and the size of the screen immediately above. $\overline{L}$ is the average of these two sizes. Needless to say, the precision of this calculation is dependent somewhat on the sample size and the number of size fractions. Usual screen analyses are adequate, but the pan fraction and the largest size fraction should not be used.

The data obtained from the calculation are then plotted on semilog paper; that is, log n vs. $\overline{L}$. The resulting plot forms a straight line with slope related to $-(1)/(rT)$ by the conversion factor from log 10 to log e. The intercept is the population density at zero size n°. Such a plot is illustrated in Figure 3. Because T is known, the growth rate r can be determined from the slope. Care must be taken in drawing the line so that the parameters n° and r which are obtained are consistent with the expression for the total mass as defined by Equation (17). The mass balance will be discussed later.

Examination of the definition of n° and r results in the following relationship:

$$n^\circ = \left.\frac{dN}{dL}\right|_{L=o} \tag{23}$$

$$r = \frac{dL}{dt} \tag{24}$$

$$\left(\left.\frac{dN}{dL}\right|_{L=o}\right)\left(\frac{dL}{dt}\right) = \left.\frac{dN}{dt}\right|_{L=o} \tag{25}$$

The right-hand member of Equation (25) is the nucleation rate; consequently, the nucleation rate is given by the product of the intercept of the distribution plot and the growth rate. The interesting feature of this situation is that both the nucleation rate and the growth rate can be obtained from the analysis of the size distribution and thus can be obtained under identical conditions.

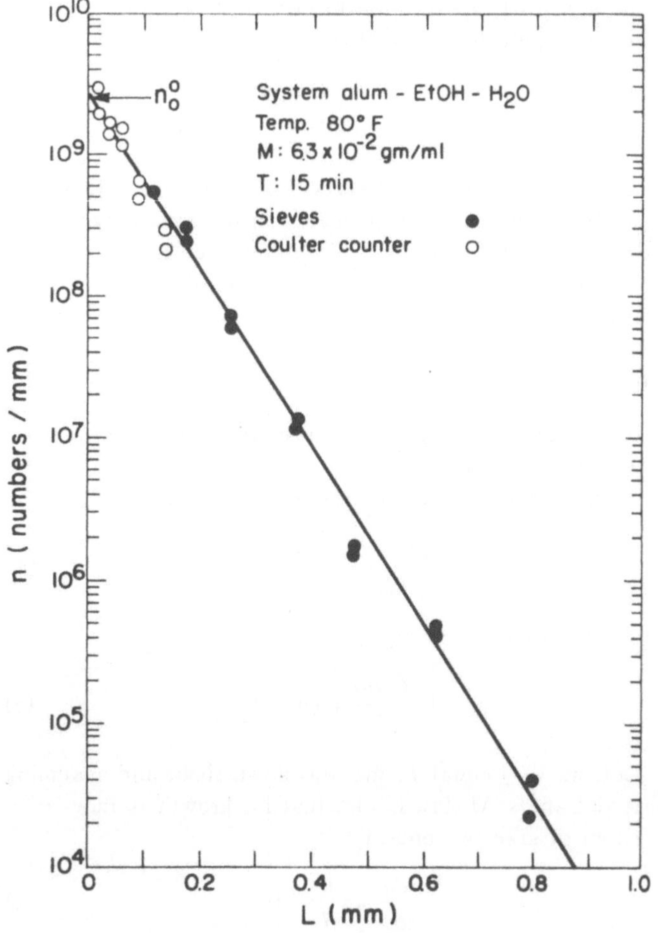

Fig. 3. Experimental size distribution.

NUCLEATION KINETICS

The above analysis suggests a convenient method for determination of the nucleation kinetics; therefore, consider now a series of experiments carried on under the constraints discussed earlier. It is possible to satisfy these constraints and conduct a series of experiments at different supersaturations but at the same suspension density. This is done most readily by using identical feeds and by maintaining the same crystallizer temperature but varying the feed rate. Changes in the feed rate change the drawdown time. Such changes in operation must result in different supersaturations if the mass balance is to be satisfied. Different supersaturations will exist, resulting in different growth and nucleation rates and, consequently, different size distributions. Such obtained size distributions can then be used to determine the kinetic rates.

If supersaturation could be precisely measured, models relating the kinetic rates to supersaturation could be developed. Supersaturation, however, is usually at such a low level that accurate measurement is nearly impossible. Useful information can still be obtained about the kinetics, even if supersaturation is impossible to measure, by use of the following development.

It has been found, generally, for mixed suspension systems that growth rate is very nearly a linear function of supersaturation:

$$r = k_g s \tag{26}$$

Furthermore, it has been shown that power law kinetic models for nucleation can be used:

$$\frac{dN}{dt} = k_n s^i \tag{27}$$

The above simplified kinetic formulation presumes a given temperature and suspension density. The adjustable parameters, of course, are the k's and the exponent i; these must be determined by experiment.

Because nucleation and growth are occuring simultaneously and in the same environment, they are dependent on the same level of supersaturation. The effect of the kinetics on the size distribution reduces then to the kinetic relationship between these two kinetic phenomena. A relationship between these rates can be obtained by combining Equations (26) and (27) to give

$$\frac{dN^\circ}{dt} = k_N r^i \tag{28}$$

or, in terms of the population density

$$n^\circ = k_N r^{i-1} \tag{29}$$

Equation (29) immediately suggests that the data obtained as described in the previous section can be used to determine k and i. Indeed, if a power law model is applicable, a log-log plot of n° and r or $(dN^\circ)/(dt)$ and r will give a straight line of slope i–1 or i, respectively. Thus, the exponent i can be determined. Such a plot of experimental data is shown in Figure 4.

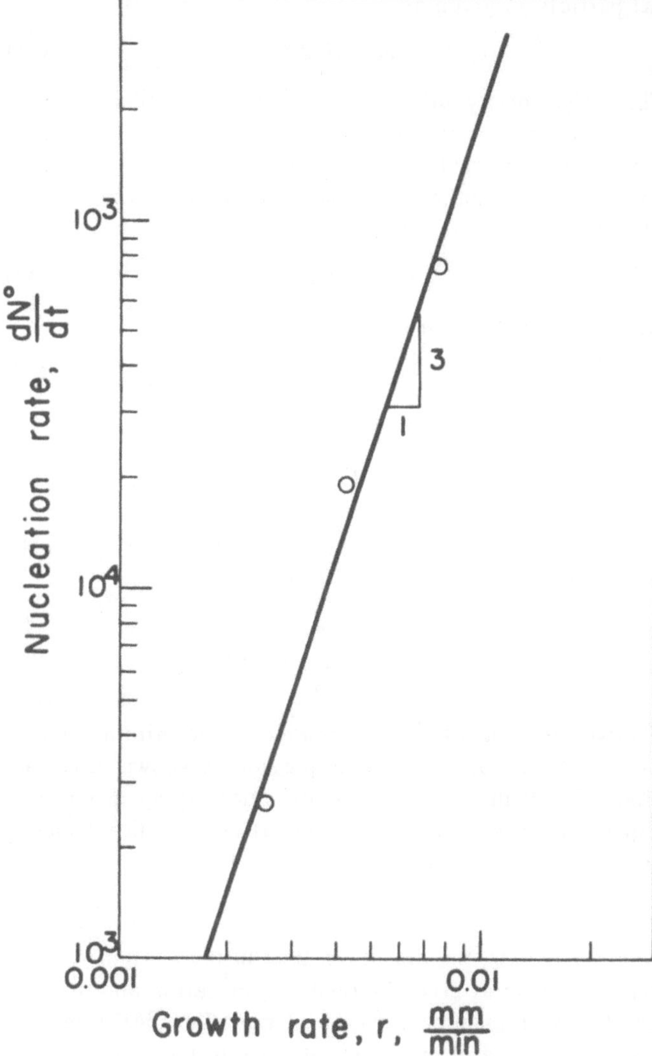

Fig. 4. Experimental nucleation kinetics.

SUMMARY

The above development provides the bases for the description of the behavior of a large class of crystallization processes and also suggests how certain kinetic characteristics can be determined. In subsequent sections the above mathematical apparatus will be used to explain the behavior of various crystallizer systems.

APPLICATIONS

MASS AND POPULATION BALANCES

The crystal size distribution (CSD) in a crystallization process, as described by a population balance, results in a total concentration of particles in the suspension, the solids concentration M_T. As the amount of solid-phase material must be taken into account in writing a mass balance for the system, then the solids concentration M_T becomes a link between the population and mass balances describing the system. These balances are shown for the single-stage mixed suspension mixed product removal (MSMPR) crystallizer and shown in Figure 2. Remember that the indicated quantities are defined as follows:

Again, by assuming that $Q_i = Q_o = Q$, the population balance becomes

$$rV\frac{dn}{dL} = -Qn \tag{5}$$

giving particle distribution

$$n = n^\circ \exp(-LQ/rV) \tag{8}$$

A mass balance on the solute gives

$$Q_i C_i = Q_o (C + M_T) \tag{30}$$

At steady state, the particle distribution $n(L)$ must satisfy both of these balances simultaneously. The solids concentration M_T can be related to the particle distribution $n(L)$ as follows. The weight of an individual particle is given as

$$m = \rho k_v L^3 \tag{31}$$

The weight of crystals per unit volume of mother liquor in the suspension in a small size range dL is equal to the weight per crystal times the number of crystals per unit volume of mother liquor dN, which are in size range dL. Thus

$$dm = \rho k_v L^3 dN \tag{32}$$

But, by the definition of population density

$$dN = n dL \tag{33}$$

Thus

$$dm = \rho k_v L^3 n dL \tag{34}$$

The total mass of particles per unit volume of liquor (solids concentration) is then the sum (integral) over all possible size ranges of dm. Thus

$$M_T = \rho k_v \int_0^\infty n L^3 dL \tag{35}$$

The solids concentration is related to the third moment of the particle distribution as previously shown in Equation (15). If the form of population density as given by Equation (8) is substituted in Equation (35), this equation can be integrated analytically. Thus

$$M_T = 6\,(rT)^4 \rho k_v n^\circ \tag{36}$$

where $T = V/Q$, the drawdown time in the system. As M_T is in general given by feed concentration and the solubility diagram of the system, Equation (36) is in effect a constraint on the growth rate of the system. For a given nuclei density, only one value of linear growth rate will satisfy the mass balance for the system.

It has been shown that the mass concentration in a given size range is

$$dm = \rho k_v L^3 n dL \tag{34}$$

The weight fraction in that size range is thus dm divided by the total concentration of particles M_T. Thus

$$dw = \frac{dm}{M_T} \tag{37}$$

Substituting the definitions of dm and M_T in Equation (35) we get

$$\frac{dw}{dL} = \frac{L^3 n}{6 n^\circ (rT)^4} \tag{38}$$

where dw/dL is the weight fraction distribution of particles in suspension. If the distribution is defined by Equation (8), the weight distribution is given as

$$\frac{dw}{dL} = \frac{\exp(-L/rT) L^3}{6\,(rT)^4} \tag{39}$$

This weight distribution, Equation (37), exhibits a maximum, which maximum occurs at the dominant size of the distribution. The dominant size can easily be found by setting the derivative of the weight distribution equal to zero. Thus

$$6(rT)^4 \frac{d}{dL}\frac{dw}{dL} = 3L^2 \exp(L/-rT) - \frac{L^3}{rT} \exp(L/-rT) = 0 \tag{40}$$

The cancelling of $L^2 \exp(-L/rT)$ leaves

$$3 - \frac{L}{rT} = 0 \tag{41}$$

or

$$L_D = 3rT \tag{42}$$

where L_D is the dominant size on a mass basis in the distribution. At first glance it appears that the dominant size is independent of the nucleation rate, which is a ridiculous situation. However, for a given M_T (given by feed concentration and solubility diagram), the growth rate obtainable depends on nuclei density as given by Equation (36). Thus growth rate, and hence dominant size, depend on nucleation rate.

SIZE DISTRIBUTION WITH COMPLEX FLOWS

Fines Removal

A common way of increasing average particle size from a crystallization process is to segregate crystals from the suspension while they are still small and destroy them, usually by steam dissolving. Actually, if the fines can be removed at a small enough size, it is immaterial whether they are dissolved or are removed with the product. The essential requirement is that they be removed with a significantly shorter drawdown time than product crystals. Of course, the smaller the size of fines that can be segregated, the more efficient will be the operation of the fines dissolver. The population density plot is an ideal way to analyze performance of such a crystallizer with a fines removal system.

Example 1:

By using the previously derived theory, construct an hypothetical semilog population density plot of CSD from a crystallizer before and after the addition of a fines removal system considered as case 2 and case 1. For this example, in case 1, it is assumed that fines, in size 0 to L_F, are withdrawn with a flow rate RQ_o, while product crystals, in size range L_F to ∞, are withdrawn at a rate Q_o, and where R is the ratio of product to fines drawdown times. Thus the drawdown times will be

$$T_F = \frac{V}{RQ_o} \text{ for } L < L_F \tag{43}$$

and

$$T_p = V/Q_o \text{ for } L > L_F$$

From previous theory it is known that crystal population in each size range will decay exponentially, inversely proportional to the product of growth rate times holding time. Thus the population densities will be of the form

$$n_F = C_1 \exp(-L/r_1 T_F) \text{ for } L < L_F \tag{44}$$

$$n_p = C_2 \exp(-L/r_1 T_p) \text{ for } L > L_F$$

where C_1 and C_2 are constants. Furthermore, $n_F = n_p$ when $L = L_F$, as the distribution is surely continuous. Thus, on the semilog plot, Figure 5, a straight line can be drawn in the segment 0 to L_F with a negative slope R times that in the range (L_F, oo) of the product crystals. The curve for the mixed product case can be located as follows. The total mass in suspension M_T must be the same in both cases 1 and 2 (with and without fines dissolving), if it is assumed that the fines are of negligible mass compared with product crystals. Thus, if both curves start out with the same nuclei density $n_1^\circ = n_2^\circ$, then curve 2 must have a slope intermediate between the slopes of the two sections of curve 1 in order for the integral of $L^3 n dL$ over the entire distribution to be equal in both cases. This locates curve 2 (qualitatively) in Figure 5. As the drawdown time for case 2 is supposedly the same as T_p in case 1, $T = VQ_o$, it must be concluded that the growth rates are different, and in fact that $r_1 > r_2$. Thus, the net effect of fines removal is to force the growth rate to a higher level, producing the same production on larger average size crystals having less total surface area.

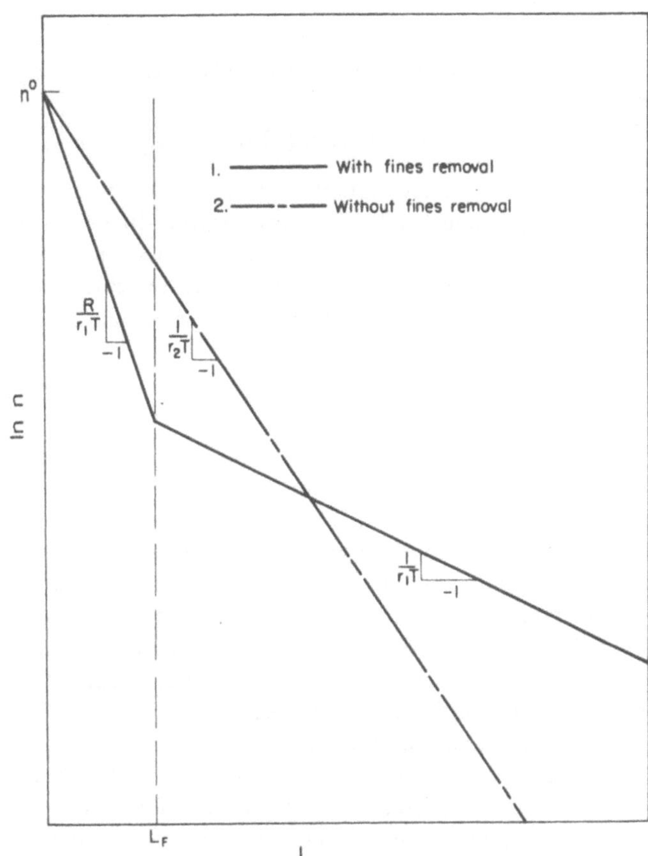

Fig. 5. Size distribution with fines removal.

The above example was worked out by assuming that the nuclei densities remained the same in both cases. In practice, this would only occur if the nucleation rate increased in the same proportion as growth rate in the case of fines removal. If the higher supersaturation produced with fines removal results in a greater than proportionate increase in nucleation rate (likely), then additional nuclei will be formed, and size improvement will be somewhat less than expected based on the previous analysis. This internal feedback in the system, which limits size improvement in regions where nucleation is a sensitive function of supersaturation, can be analyzed with an idealized model of the fines removal system. Thus, assume an idealized system, given as follows:

1. Power model nucleation-growth rate kinetics of the form $n^\circ = k_N r^{i-1}$.
2. Fines segregated at the limit of zero size, that is, true nuclei dissolving.
3. Fines destroyed proportional to their population $n_1^\circ = \beta n^\circ$.

Under these limiting assumptions, it can be shown that the expected dominant size with fines dissolving is given as

$$L_{D2} = L_{D1} \left(\frac{1}{\beta}\right)^{\frac{1}{i+3}} \tag{45}$$

Thus, as the relative sensitivity of nucleation to growth increases, given by an increase in the parameter i, less size improvement is achieved with a given fraction of fines removal $1 - \beta$.

A real crystallizer with fines removal can be analyzed rigorously, without the previous idealizations, by writing population and mass balances, together with nucleation-growth rate kinetics. Such rigorous analysis depends on accurate knowledge of actual crystal drawdown rates as a function of size as well as nucleation-growth rate kinetics in the existing range of supersaturation. In general, computer solution of the equations would be required.

Classified Product Removal

As a mixed suspension crystallizer produces an inherently wide distribution of particle sizes, one is tempted to say, "let's leave all the crystals in the suspension until they are product size, say L_p, and then remove them immediately." This case, representing ideal classification, has been thoroughly studied in comparison with the mixed product removal case and in theory would produce monosized crystals. One very interesting result arising from the comparison of these two cases, mixed and perfectly classified removal, is that under comparable operating conditions the mixed product removal case requires approximately eleven times as many nuclei crystals to produce the same production rate with an average size L_p equal to the classified product crystals. It would be better to say that the classified case can tolerate only one-eleventh the nucleation rate as in the mixed case and still produce the same average size. Thus, product classification is almost always accompanied by a reduction in average size, unless nucleation is independently controlled. In the mixed withdrawal case, many small crystals are withdrawn in the product before they are of appreciable size. Consequently, mixed withdrawal is an effective means of population control.

The concept of the ideal classified product removal crystallizer as outlined above is not too realistic when applied to practice, mainly because of the following:

1. In crystallizers which require withdrawal of mother liquor with product crystals to maintain overall material balance control, it is impossible to remove this mother liquor without removing a proportionate amount of undersize particles. This is because classifying devices (for example, hydroclones, screens, etc.) exclude larger crystals from the overflow recycle, while undersize particles split with nearly the same ratio as liquor overflow to underflow. In the case of a cooling crystallizer, this minimum underflow discharge of mother liquor is equal to normal mixed product discharge rate. Thus, undersize crystals are in no sense recycled as the model suggests, but instead, removal of oversize crystals is enhanced. In the case of an evaporative crystallizer, liquor withdrawal rate is arbitrary (withdrawal rate controls solids concentration M_T), and very low withdrawal rates of undersize crystals can be achieved. This is also true of adiabatic flash crystallizers where two compounds react to form a crystal product with only enough water added to remove heat of reaction. An example of the latter case is ammonium sulfate produced by reacting ammonia with fairly concentrated sulfuric acid solutions, having just enough water to remove the heat of reaction by flashing. In such a case, it would be theoretically possible to remove no liquor from the process and hence to totally recycle undersize crystals.

2. In order to immediately remove oversize particles after they grew to size L_p, one would need an infinitely large flow rate to a classifying device. As practical recycle rates might be in the range 1.5 to 5, this limits the rate of removal of oversize. In practice, crystals above the product size are removed at a faster than normal rate, producing a narrower, but not monosized, crystal distribution.

As with the case of fines removal, a classified product removal crystallizer can be rigorously analyzed by simultaneous solution of population and mass balances together with nucleation-growth rate kinetics. Again, accurate data concerning actual drawdown rates as a function of crystal size are required. The following example illustrates an idealized, but more realistic, model of a classified product removal crystallizer. The model describes the real situation much better in cases where large amounts of mother liquor must be withdrawn to maintain the overall material balance.

Example 2:

By using the previously derived theory, construct a hypothetical semilog population density plot of CSD from a cooling crystallizer before and after installation of a classifying device, say a hydroclone, classifying at a particle size L_p. These two situations, before and after addition of classifier, will be referred to as case 1 and case 2. Again, it is assumed that any changes in supersaturation brought about by classification result in proportionate changes in nucleation and growth rate so that $n_1^\circ = n_2^\circ$. The classification at size L_p is assumed to be sharp, such that the idealized undersize and oversize particle drawdown times can be defined as

$$T_u = V/Q_o \text{ for } L < L_p \tag{46}$$

$$T_o = V/zQ_o \text{ for } L > L_p$$

The population density in the two size ranges is then given as

$$n = C_1 \exp(-L/r_2 T_u) \text{ for } L < L_p \tag{47}$$

and

$$n = C_2 \exp(-L/r_2 T_o) \text{ for } L < L_p$$

where C_1 and C_2 are constants. As the population density must be continuous at $L = L_p$, two straight-line segments shown in Figure 6 can be drawn with negative slopes $1/r_2 T$ and $z/r_2 T$ below and above the classification size L_p. The mixed product removal crystallizer, case 1, is given by

$$n = n^\circ \exp(-L/r_1 T) \tag{48}$$

In this example, the integral of $L^3 ndL$ over the entire distribution is not the same for both cases but is less for the case of classification, as the slurry density has been drawn down by the preferential removal of oversize crystals. Thus, to produce the same production, the growth rate is forced to a higher level, and it is concluded that $r_2 > r_1$. Hence the curve for the classified case (for $L < L_p$) decreases with a smaller negative slope than the mixed removal case.

In the general case, nucleation will not increase in the same ratio as growth rate, as assumed in the above example, and the expected CSD with product classification can only be calculated by simultaneous solution of population and mass balances with nucleation-growth rate kinetics.

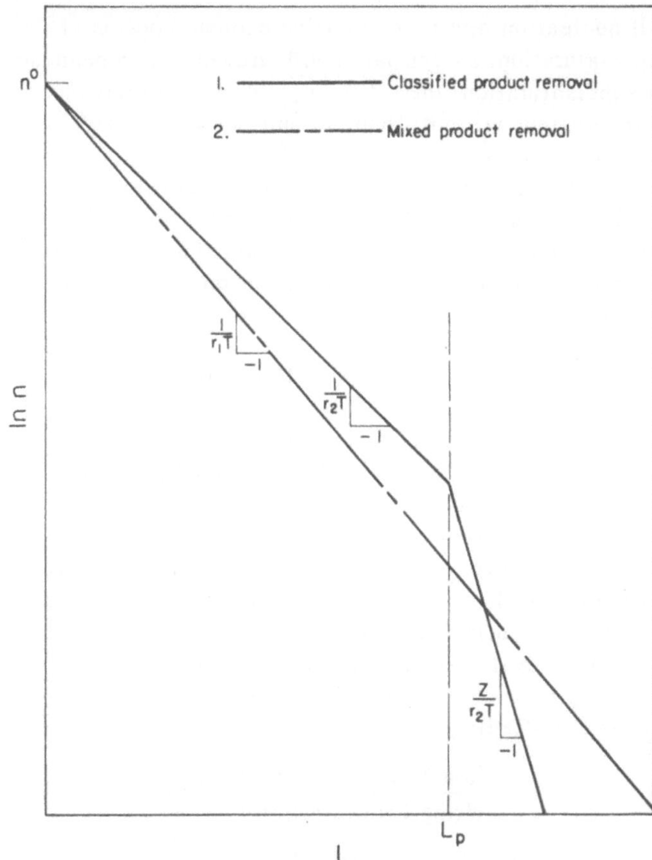

Fig. 6. Size distribution with classified product.

Arbitrary Solids-Retention Crystallizer

An ideal crystallizer would permit the separation, at will, of residence-time distributions of solids vs. liquor and solids as a function of size. Thus, one could operate with arbitrary solids concentrations (solids vs. liquor separation) and produce a narrow CSD with classified removal while keeping nucleation in control with a fines destruction system. Low driving forces (supersaturation) would be maintained by recycling a heavy seed bed intermediate in size between fines and product L_F and L_P. Although any equipment vendor would sell such a crystallizer on demand, it is doubtful if such a versatile machine exists, mainly because it would take a very cooperative crystal system to allow the classification implied in such a process. There is a very strong interaction between the crystal system and design of a suitable machine for crystallization, which fact has given rise to much of the art in crystallizer design.

The following example considers an ideal arbitrary residence time crystallizer by using the previously developed theory for analysis for CSD.

Example 3:

By using the previously derived theory, construct a hypothetical semilog population density plot of CSD from a crystallizer with clear liquor removal, fines destruction system, and classified product removal. Use the same idealizations as in the previous examples.

1. Clear liquor overflow Q_o separates the residence time of solids relative to liquor, thus affecting the total solids concentration. However, the form of CSD is not affected. This clear liquor overflow must, of course, be taken into account in writing mass balances, but not population balances. The independent control of M_T with clear liquor overflow gives the adiabatic or cooling crystallizer the same versatility on product drawdown as that of the evaporative crystallizer. Net liquor withdrawal with classified product crystals determines drawdown rate of crystal seed bed.

2. Fines destruction will give a large drawdown rate for crystals below size L_F. As a result a plot of the fines size distribution will have a slope R times that of the crystal seed bed. This is shown by the first straight-line segment in Figure 7. The second segment, representing the crystal seed bed, greater in size than fines but less than product crystals, decays with a negative slope of $1/rT$, where $T = V/Q_u$ and where Q_u, the underflow discharge rate, is determined by the amount of clear liquor overflow Q_o.

3. Classified product removal above product size L_p will give a drawdown in this size range z times that of the seed bed drawdown. This gives the final straight-line segment in Figure 7 with a negative slope of z/rT.

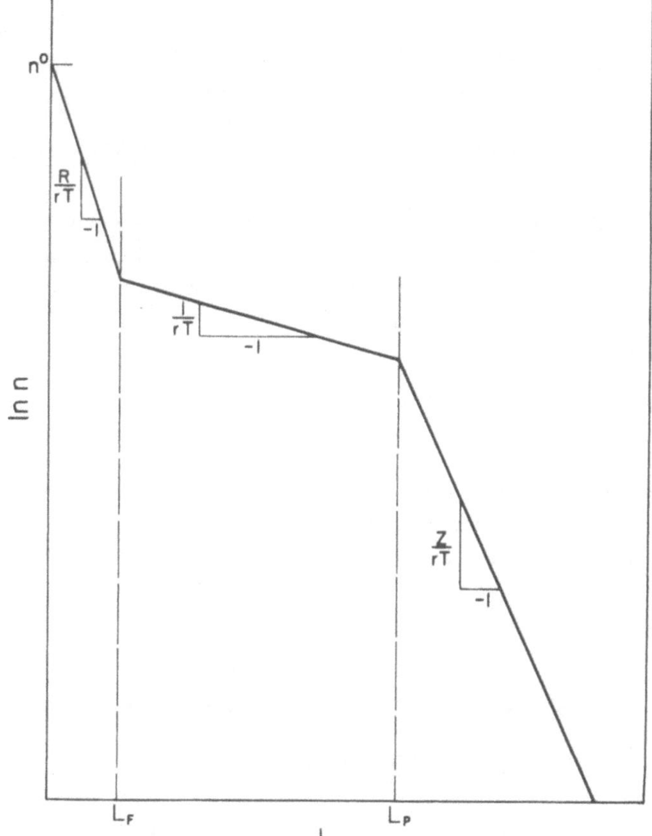

Fig. 7. Combined fines removal and classified product size distribution.

The resultant solids concentration M_T and crystal growth rate r can be found by simultaneous solution of population and mass balances together with nucleation-growth rate kinetics.

As in previous examples, rigorous analysis of the arbitrary residence time crystallizer can be made if actual drawdown rates as a function of size as well as accurate nucleation-growth rate kinetics in the resulting region of supersaturation are known for the system. It should be obvious from all of these examples that much remains to be learned about actual classification mechanics as well as nucleation-growth rate kinetics.

Stability of CSD

In the original derivation of the population balance, accumulation was not considered because steady state only was considered. In order to consider stability problems, the unsteady state must be considered:

$$\text{Accumulation} = \text{Input} - \text{Output} + \text{Generation} \qquad (49)$$

At steady state and with no breakage, the accumulation and generation terms are zero, but under transient conditions, an accumulation of crystals in a size range ΔL is possible. The rate of accumulation in size range is given by

$$\frac{\partial}{\partial t}(\Delta N) = \frac{\partial}{\partial t}(Vn\Delta L) \qquad (50)$$

Putting this rate of accumulation term in the previous steady state population balance, we get (Figure 2)

$$\frac{\partial}{\partial t}(Vn\Delta L) = Vnr|_L - Vnr|_{L+\Delta L} - nQ\Delta L \qquad (51)$$

Dividing by $V\Delta L$ and taking the limit as $\Delta L \longrightarrow 0$ we obtain the transient form of the population balance which, together with appropriate initial and boundary conditions, describes CSD transients in a mixed crystal suspension. Thus

$$\frac{\partial n}{\partial t} = -r\frac{\partial n}{\partial L} - \frac{n}{T} \qquad (52)$$

where again $T = V/Q$. In addition to Equation (52), it is necessary to specify an initial particle distribution, nucleation rates, and growth rates as a function of time in order to solve Equation (52) to describe transients after a given upset. These additional equations can be written symbolically as

Initial distribution $\quad n(o,L) = g(L) \qquad (53)$

Nucleation-growth rate kinetics $\quad n(t,0) = h(r) \qquad (54)$

Constraint on growth rate such that production rate is maintained $\quad r \alpha \dfrac{\text{Production rate}}{\text{Total crystal area}} \qquad (55)$

If nucleation rate is a sensitive enough function of supersaturation, as compared with growth rate dependence on supersaturation, then it is possible that disturbances to the system will not damp out and the system will exhibit a sustained cycling of CSD, that is, become unstable. Figure 8 illustrates this basic feedback in the system, which in theory can produce instability. The essence of this figure is that the molecular driving force (supersaturation) is influenced by the total surface area available for growth, given by CSD at a given instant. But nucleation and growth rates depend on supersaturation and in turn determine CSD. Hence the internal feedback as it is shown in Figure 8.

Equation (52) can be used to analyze stability as well as transients of CSD. To study stability, Equation (52) is first transformed to a set of ordinary, but nonlinear, differential equations in terms of the moments of the distribution. These moment equations are then linearized and their stability investigated by classical root locus techniques. Such studies have indicated that the mixed product removal crystallizer will be stable as long as the following relationship between nucleation rate and growth rate is satisfied:

$$\frac{d(\log \text{Nucleation Rate})}{d(\log \text{Growth Rate})} < 21 \qquad (56)$$

The assumption of power model kinetics would indicate that $i < 21$; for homogeneous nucleation as a function of supersaturation, this means that nucleation would have to have twenty-one times the kinetic order (in terms of supersaturation) as the kinetic order of growth rate in order for the system to become unstable. Such CSD instability in a mixed product crystallizer would require essentially a discontinuity in the nucleation rate, for example, passing a metastable boundary. Theoretically proposed models for homogeneous nucleation, which describe nucleation as an all-or-nothing function of supersaturation, exhibit rates of change near the critical supersaturation such that inequality (56) is violated and instability could exist. On the other hand, much kinetic data gathered in real operating crystallizers indicate a rather gentle function of nucleation rate vs. growth rate throughout most of the operating range, indicating that

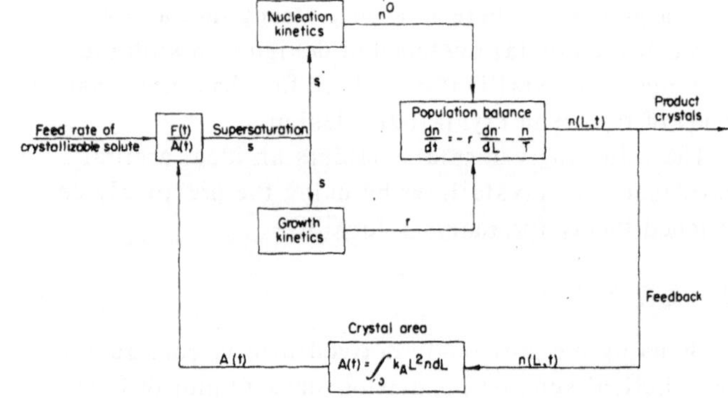

Fig. 8. Feedback effects in continuous crystallization.

the mechanism for nucleation involves heterogeneous factors and is not purely homogeneous. Thus, a reasonable explanation for some cases of observed CSD instability is that the driving forces were increased (usually by fines dissolving) to a metastable boundary where the homogeneous mechanism created an essentially discontinuous increase in total nucleation.

Studies have also been made of CSD stability by using the model of classified product removal which assumes total recycle of particles below the classification size L_p and instantaneous removal above size L_p. In this case, the stability criterion, Equation (54), was relaxed to more reasonable values (about two to five, depending on withdrawal rate) such that stability might be expected with heterogeneous nucleation kinetics. However, this classification model is very unrealistic except in the case where no mother liquor is removed from the system. It can be shown that classified product removal, given by the model described in this paper (where the removal rate of undersize is unchanged but the removal rate of oversize is enhanced), actually stabilizes CSD.

It is quite possible that all cases of CSD instability, with mixed or classified withdrawal, are due to operating conditions which force the system past a metastable boundary. It should be noted that this could be a lower as well as upper boundary, for example, a low level of supersaturation where the supersaturation induced heterogeneous mechanism ceased, and the only source of nuclei was due to attrition and perhaps external seed.

The question of instability of CSD is far from settled, but the method of describing CSD with transient population and mass balances together with nucleation-growth rate kinetics provides a rigorous tool for the analysis of such systems.

ACKNOWLEDGMENT

The authors wish to acknowledge the assistance of Dr. John D. Stevens of Iowa State University for his careful review of this paper and for his assistance in its original workshop presentation at Tampa, Florida, in May, 1968.

NOTATION

A = crystal area
C = solute concentration
C_i = feed concentration of solute, lb./cu. ft.
C_o = concentration of solute in crystallizer and discharge, lb./cu. ft.
i = nucleation order
k_A = area shape factor
k_g = kinetic constant for growth
k_n = kinetic constant for nucleation
$k_N = k_n/k_g^{\,i}$
k_v = volume shape factor
L = crystal size, μ
$\bar{L}$ = arithmetic average particle size in fraction
$\mathfrak{L}$ = total crystal length
m = mass
M = suspension density
M_T = solids concentration, lb./cu. ft. (clear liquor basis)
n = population density [number/(unit length) (unit slurry volume)]
$\bar{n}$ = average population density in size interval
n^o = nuclei population density, No./(μ) (cu. ft.)
N = number of crystals
Q = volumetric feed and discharge rate
Q_i = feed flow rate, cu. ft./sec.
Q_o = discharge flow rate, cu. ft./sec. (clear liquor basis)
r = linear particle growth rate, ft./sec.
R = ratio of fines residence time to coarse residence time
s = supersaturation
t = time
T = drawdown time
V = volume of suspension, cu. ft.
w = mass fraction
x = dimensionless size L/rT
z = feed to underflow ratio
β = fraction of nuclei remaining
ΔL = width of crystal size range
ρ = particle density

SUBSCRIPTS

T = total
F = fines
p = product
o = overflow
u = underflow

SUGGESTED REFERENCES

POPULATION BALANCES

1. Hulburt, H. M., and S. Katz, *Chem. Eng. Sci.*, **19**, 555 (1964).
2. Randolph, A. D., *Can. J. Chem. Eng.*, **42**, 280 (1964).

NUCLEATION-GROWTH RATE KINETICS

1. Amin, A. B., and M. A. Larson, *Ind. Eng. Chem. Process Design Quart.*, **1**, 133 (1968),
2. Bransom, S. H., W. J. Dunning, and B. Millard, *Disc. Faraday Soc.*, **5**, 83 (1949).
3. Cayey, N. W., and Joseph Estrin, *Ind. Eng. Chem. Fundamentals*, **6**, 13 (1967).
4. Larson, M. A., D. C. Timm, and P. R. Wolff, *AIChEJ.*, **14**, No. 3, p. 448 (1968).
5. Murray, D. C., and M. A. Larson, *ibid.*, **11**, 728 (1965).
6. Nancollas, G. H., and N. Purdie, *Chem. Soc. Quart. Rev. (London)*, **18**,(1964).
7. Timm, D. C., and M. A. Larson, *AIChEJ*,, **14**, No. 3, p. 452 (1968).

CRYSTAL SIZE DISTRIBUTION STUDIES

1. Abegg, C. F., J. D. Stevens, and M. A. Larson, *AIChEJ.*, **14**, No. 1, p. 118 (1968).
2. Bennett, R. C., *Chem. Eng. Progr.*, **58**, 76 (1962).
3. Bransom, S. H., *Brit. Chem. Eng.*, **5**, 838 (1960).
4. Canning, T. F., and A. D. Randolph, *AIChEJ.*, **13**, 5 (1967).
5. Randolph, A. D., and M. A. Larson, *ibid.*, **8**, 639 (1962).
6. Randolph, A. D., *ibid.*, **11**, 424 (1965).
7. Robinson, J. N., and J. E. Roberts, *Can. J. Chem. Eng.*, **35**, 105 (1957).
8. Saeman, W. C., *AIChEJ.*, **2**, 107 (1956).

DYNAMICS AND STABILITY OF CSD

1. Han, C. D., and R. Shinnar, *AIChEJ.*, **14**, No. 4, 612 (July, 1968).
2. Randolph, A. D., and M. A. Larson, *ibid.*, **8**, 639 (1962).
3. ———, *Chem. Eng. Progr. Symposium Ser. No. 55*, **61**, 147 (1965).
4. Sherwin, M. B., R. Shinnar, and S. Katz, *AIChEJ.*, **13**, 1141 (1967).
5. ———, paper presented at AIChE 61 National Meeting, Houston, Tex. (1967). Stirred Crystallizer with Classified Product Removal. 61st National Meeting AIChE, Houston, Texas, Paper 12F.

SAMPLE PROBLEMS: POPULATION BALANCE

PROBLEM 1

Sieve analysis data are given in Table 1 from a series of runs carried on in a CMSMPR crystallizer. The feeds for runs I, II, and III were identical, and the measured suspension densities were approximately the same for each run. The only constraint which was varied was the drawdown time.

The population densities have been computed for run I and plotted on the accompanying semilog plot. The results of run III have also been plotted, but the data have not been given. For run II only the screen analysis has been given.

1. Compute the distribution for run II in terms of population density and plot on the same graph as the other two runs. [Be sure that your line gives values for r and n° which satisfy $M_T = 6k_v n^\circ (rT)^4$.]

2. Determine r and n° for all three runs and plot n° vs. r on log log paper. Determine a power law kinetic model.

3. (Optional. Plot the population densities for run I and see if the points fall on the line given. Justify the use of the straight line in light of the fact that some points do not lie on the line.

4. (Optional). Determine the mass distribution from the plot for run III.

A partial solution is shown in Figure 9.

TABLE 1. SIZE DISTRIBUTION FROM CMSMPR CRYSTALLIZATION

Standard U.S. Sieve	L, μ Ave. Size	ΔL μ	$\rho k_v \bar{L}^3 \Delta L$	I Q = 3.6 l/min T = 15 min		II Q = 1.8 l/min T = 30 min		III Q = 0.9 l/min T = 45 min	
				w,*gm	n	w,*gm	n	w,*gm	n
+35				1.30					
-35 + 40	479	66	1.28×10^{-2}	0.76	59	0.10			
-40 + 50	374	145	1.34×10^{-2}	1.70	127	1.03			
-50 + 70	257	89	2.68×10^{-3}	3.82	1,425	8.68			
-70 + 100	177	71	6.98×10^{-4}	9.03	12,940	26.18			
-100 + 140	128	26	9.65×10^{-5}	12.72	132,000	25.58			
-140 + 200	98	34	5.65×10^{-5}	27.10	480,000	28.92			
-200 + 325	63	37	1.64×10^{-5}	44.70	2,730,000	17.48			
-325				10.72		3.06			
TOTAL				111.85		110.97		110.00	

$n = \dfrac{w}{\rho k_v \bar{L}^3 \Delta L}$ $\rho = 1.77 \times 10^{-12}$ gm/μ^3

$k_v = 1$ * Solids in 1 liter of slurry

Note: See plots

SAMPLE PROBLEMS: APPLICATIONS

1. Estimate the fraction of nuclei that would have to be destroyed to produce a 10% increase in dominate crystal size in the following kinetic regions: a metastable region where the ratio of kinetic order of nucleation to growth i was zero, and a labile region where $i = 9$.

2. An MSMPR$_7$ crystallizer operates with a steady state value of 1.0×10^7 nuclei/cc.-cm., a growth rate of 0.13 μ/min., and mixed product withdrawal rate (clear liquor basis) of 9.0 gal./min. The crystallizer volume is 4,750 gal. [clear liquor basis, the crystal density is 1.93 g./cc., and the volumetric shape factor is 0.6 cu. ft./(length)^3]. What is the solids concentration in the crystallizer, grams per cubic centimeter, and what is the production rate of crystals, pounds per hour.

3. In the above example (2), what fraction of original nuclei have been removed in the discharge by the time the nuclei have grown to 50 μ.

Assume that a fines removal system in the size range 0 to 50 μ is installed, which removes fines at a rate Q_F gallons per minute in addition to the 9.0 gal./min. product rate. Neglecting any changes in nuclei density and growth rate due to the fines removal system, calculate the fines removal rate, Q_F gallons per minute, necessary to remove nine out of ten original nuclei by the time they have grown to 50 μ in size.

ANSWERS TO PROBLEMS ON APPLICATIONS

1. By using the idealized model of a fines dissolver

$$L_{D_2}/L_{D_1} = \left(\frac{1}{\beta}\right)^{\frac{1}{i+3}}$$

$$L_{D_2}/L_{D_1} = 1.10/1.0 = 1.1$$

Thus

$$\frac{1}{\beta} = (1.1)^{i+3}$$

or

$$\beta = \frac{1}{(1.1)^{i+3}}$$

For

$$i = 0 ,$$

$$\beta = \frac{1}{(1.1)^3} = 0.752$$

Thus 1.0 to 0.752 or 24.8% of the nuclei must be destroyed in the metastable region.

If $i = 9$

then $\beta = \dfrac{1}{(1.1)^{12}} = \dfrac{1}{3.14} = 0.318$

Thus 68.2% of the nuclei must be destroyed in the labile region.

2. $M_T = 6\rho k_v n^\circ (rT)^4$

$$rT = \frac{(0.13)(4,750)}{9.0}$$

$$= 68.7\mu$$

$$= 68.7 \times 10^{-3} \text{ cm.}$$

$$(rT)^4 = 2.23 \times 10^{-9} \text{ cm.}^4$$

$$M_T = (3.6)(1.93)(2.23)(10^7)(10^{-9})$$

$$= 0.155 \text{ g./cc.}$$

Production P is the product of solids concentration times the flow rate.

$$P = M_T Q_o$$

$$M_T = (0.155)(62.3) = 9.67 \text{ lb./cu. ft.}$$

$$Q = 9.0 \text{ gal./min.} = \frac{(9.0)(60)}{7.68} = 72.3 \text{ cu. ft./hr.}$$

$$P = (72.3)(9.67) = 698 \text{ lb./hr.}$$

3. Population decays exponentially with size. Thus

$$n/n^\circ = \exp(-L/rT)$$

$$rT = 68.7\mu \text{ for problem 1}$$

$$x = \frac{L}{rT} = \frac{50}{68.7} = 0.728$$

$$n/n^\circ = \frac{1}{e^{0.728}} = \frac{1}{2.07} = 0.483$$

Thus 51.7% of the particles **have** been removed at 50μ.

If nine out of ten nuclei have been removed, then $n/n^\circ = 0.1 = e^{-x}$.

$$x = \ln 10 = 2.303$$

$$x = \frac{L}{rT} = \frac{50}{0.13T} = 2.303$$

$$T = \frac{50}{(0.13)(2.303)} = 167 \text{ min.}$$

$$T = \frac{V}{Q_o + Q_F} = \frac{4,750}{Q_F + 9.0} = 167$$

$$Q_F = \frac{4,750}{167} - 9.0 = 19.4 \text{ gal./min.}$$

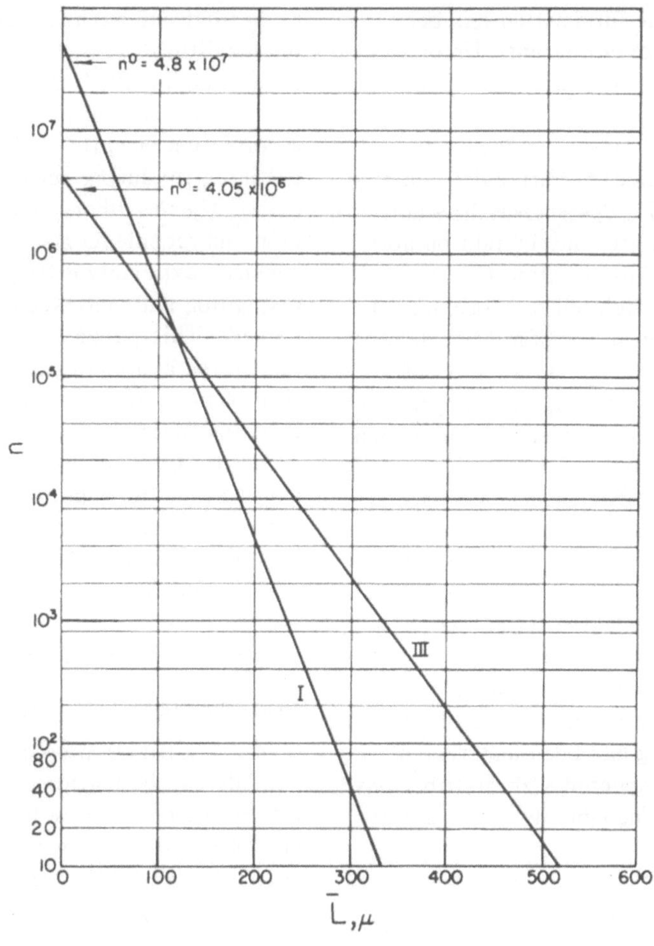

Fig. 9. Partial solution to problem 1.

MIXING EFFECTS IN CONTINUOUS CRYSTALLIZATION

G. W. Becker, Jr.
and
M. A. Larson

It is well known that idealized backmixed or plug flow conditions are seldom achieved in large industrial crystallizers. As a consequence, predicted size distributions are seldom achieved. This paper discusses the effect of various mixing concepts and process geometries on expected size distribution. Models assuming combinations of well-mixed and plug flow sections are discussed, as well as segregated flow models.

Factors influencing size distribution in well-backmixed crystallizers have been extensively studied with considerable success. Valuable information has been found concerning production size equipment. Unfortunately, in large systems good backmixing is seldom if ever achieved. In fact, it is often purposely avoided, and as a result improvements in size distribution are achieved. Therefore, the question that arises is can any significant analysis be made of a system which is not perfectly backmixed which will give any insight on how the mixing conditions effect the distribution (*1*).

The concept of well mixed as related to poorly mixed is an illusive one and has not been satisfactorily developed even for homogeneous systems. As a matter of fact, the problems related to poor mixing in homogeneous reacting systems where the reaction order is greater than 1 are intractable. Consideration of crystallization is even more complicated than nonlinear homogeneous reaction systems because of the following:

1. The degree of mixing of two phases must be considered.
2. A distribution of the solid phase must be considered.
3. The kinetics are almost certainly nonlinear.

In this discussion the consideration of some limiting conceptual cases of poor mixing has been undertaken. The study endeavors to show expected size distributions from such mixing conditions. Undoubtedly, there are no real systems which exhibit exactly the assumed mixing conditions, but the results should give some guidelines as to what one might expect from some intermediate mixing condition.

Levenspiel (*2*), Curl (*3*), and Shinnar (*4*) have considered the concept of a macrofluid. This concept assumes a collection of balls of fluid which maintain their physical integrity during some portion of their stay in a process. These authors considered in general the dispersion and recoalescence of these balls. The problem becomes extremely difficult to solve under conditions of dispersion and coalescence when crystallization is concerned. This paper considers the limiting case when no dispersion and coalescence take place, but the residence time of the balls is defined by some residence time distribution function.

An alternate approach to poor mixing problems is to use the classic model of a plug flow section in series with a well-backmixed section. This concept has also been considered and size distribution determined for certain arbitrary parameters.

RESEARCH EFFORT

NUCLEATION AND GROWTH RATE

In crystallization processes two kinetic rates must be considered, nucleation rate and growth rate. Nucleation rate is the rate at which nuclei are formed per unit volume of solution. Growth rate is the rate at which a characteristic length of a crystal elongates.

Generally, growth rate and nucleation rate are func-

Iowa State University, Ames, Iowa.

tions of the level of saturation. Neilsen (5) noted that a simple power model in the level of saturation suffices over a wide range of levels of saturation for both nucleation rate and growth rate in many systems. For continuous crystallizers, growth rate and nucleation rate have been correlated for various systems by the following relations:

$$r = k_g (C - C_{eq}) \tag{1}$$

$$\frac{dN^\circ}{dt} = k_n (C - C_{eq})^i \tag{2}$$

For some systems, growth rate has been found to be a function of particle size (6) as well as the level of agitation (1). In addition, nucleation rate is a function of the level of agitation and the suspension density (7, 8) for some systems. In this paper, however, only the simple power models, Equations (1) and (2), will be considered.

PARTICLE BALANCE

The following derivation is essentially a modified form of the balance made by Randolph and Larson (9). The crystal or particle balance is made with the aid of a population density function which is defined by the following relations:

$\int_{L_1}^{L_2} n(L, \bar{x}, t)\, dL$ = total number of particles in the size range (L_1, L_2) at time t, at position $\bar{x}$, per volume of solution

$\bar{x}$ = coordinates of space

$n(L, \bar{x}, t)$ = point population density function

L, t = characteristic length, time.

We will consider only crystals in the fixed differential size range $(L, L + dL)$ and construct a numbers balance over a fixed volume V bounded by a stationary surface with outward normal $\vec{N}_s$. In general, portions of the surface enclosing V will have fluid and crystals passing through it, and these portions of the entire bounding surface will be denoted by S. Assuming that crystals enter any arbitrary size interval only by growth, one can write the numbers balance for the crystals in $(L, L + dL)$ as

$$\frac{d}{dt}\int_V n\, dL\, dV = -\int_S n\vec{v}\cdot\vec{N}_s\, dL\, dS + \int_V (nr|_L - nr|_{L+dL})\, dV \tag{3}$$

or by eliminating dL and rearranging

$$\int_V \left[\frac{\partial n}{\partial t} + \frac{\partial}{\partial L}(nr)\right] dV = -\int_S n\vec{v}\cdot\vec{N}_s\, dS \tag{3a}$$

or

$$\int_V \left[\frac{\partial n}{\partial t} + \frac{\partial}{\partial L}(nr) + \vec{\nabla}\cdot n\vec{v}\right] dV = 0 \tag{3b}$$

The term $\vec{v}$ is the velocity of a crystal of size L at position $\bar{x}$ and time t. If one focuses his attention within a crystallizing suspension, it is obvious that V may be chosen arbitrarily, and hence the argument of Equation (3b) is identically zero. Thus, a general numbers balance equation is obtained:

$$\frac{\partial n}{\partial t} + \frac{\partial}{\partial L}(nr) + \bar{\nabla}\cdot n\bar{v} = 0 \tag{4}$$

Numbers balances for idealized flow patterns can be obtained from Equations (3a) and (4). For instance, a plug flow numbers balance is obtained from Equation (4):

$$\frac{\partial n}{\partial t} + \frac{\partial}{\partial L}(nr) = -v_x \frac{\partial n}{\partial x} \tag{5}$$

The backmix crystallization numbers balance is obtained from Equation (3a):

$$\frac{\partial n}{\partial t} + \frac{\partial}{\partial L}(nr) = \frac{n_i}{\tau_i} - \frac{n}{\tau_0} \tag{6}$$

The batch crystallizer numbers balance is a special case of Equation (6) and is given by

$$\frac{\partial n}{\partial t} + \frac{\partial}{\partial L}(nr) = 0 \tag{7}$$

It should be noted that Equations (6) and (7) are for constant volume systems. Equations (5), (6), and (7) will be utilized in subsequent sections.

MOMENT EQUATIONS

When we consider particle distributions, it is helpful to deal with the moments of that distribution. Moments describe the average properties of a particle distribution and are needed in the general solution of Equations (5), (6), and (7).

The jth moment of the population density function n is defined as

$$\mu_j = \int_0^\infty n L^j\, dL$$

With the appropriate geometric factors accounting for the crystal's shape, it is possible to determine the total numbers, length, area, volume, and mass per unit volume of solution; that is

μ_0 = Total No. $\quad K_A \mu_2$ = Total A. $\quad \rho_c K_v \mu_3$ = Total M.
μ_1 = Total Len. $\quad K_v \mu_3$ = Total V.

The unsteady state moment equations are obtained by taking the jth moment of the unsteady state particle balance describing the crystallization setup. For instance, for an ideal batch reactor, Equation (7), the moment equations are found by the operation

$$\int_0^\infty L^j \frac{\partial n}{\partial t}\, dL + \int_0^\infty L^j \frac{\partial (nr)}{\partial L}\, dL = 0$$

If one takes the size of nuclei as zero and considers the case when growth rate is not a function of crystal size, then

$$\frac{d\mu_j}{dt} = n_0^\circ\, r\,(0)^j + j\, r\, \mu_{j-1} \tag{8}$$

The first term on the left side of Equation (8) evaluated at j equal to zero is equal to the nucleation rate; that is

$$n_0^\circ\, r = \frac{dN_{(0)}}{dL}\,\frac{dL}{dt} = \frac{dN^\circ}{dt}$$

and from Equations (1) and (2) it can be seen that

$$\frac{dN^\circ}{dt} = \frac{k_n}{k_g^i}\, r^i \qquad n^\circ = \frac{k_n}{k_g^i}\, r^{i-1}$$

Finally, the mass balance for a batch reactor is

$$\frac{dC}{dt} = -\frac{d\mu_3}{dt}\, K_V\, \frac{(\rho_c - C)}{(1 - K_V\, \mu_3)} \tag{9}$$

The set of Equations (7), (8), and (9) describes the behavior of a batch crystallization. This set of equations is the same as that describing the ideal plug flow reactor if one changes variables by substituting x/v_x for t.

It is assumed that the volume occupied by the crystals is small compared with the volume occupied by the bulk solution or ($K_V \mu_3 << 1$) and ($\rho_c >> C$). With these assumptions, the set of Equations (7), (8), and (9) is reduced to the following set of equations:

$$\frac{\partial n}{\partial t} + r\,\frac{\partial n}{\partial L} = 0 \tag{10}$$

$$\begin{aligned} \frac{d\mu_0}{dt} &= K_1 r^i \qquad K_1 = k_n/k_g^i \\ \frac{d\mu_1}{dt} &= r\,\mu_0 \\ \frac{d\mu_2}{dt} &= 2\, r\, \mu_1 \\ \frac{dr}{dt} &= -3\, K_2\, r\, \mu_2; \qquad K_2 = k_g K_v \rho_c. \end{aligned} \tag{11}$$

MACROMIXING

The population density for an ideally mixed or backmix crystallizer is well known. Specifically, when the particle growth rate and the nucleation rate are given by Equations (1) and (2), the population density and the growth rate are related by the equations

$$\begin{aligned} n\,(L) &= n_0^\circ\, e^{-L/r\tau} \\ n_0^\circ &= K_1\, r^{i-1} \end{aligned} \tag{12}$$

These equations have been used by Larson and others (*9, 10*) to obtain the parameters i, n°, and r for several crystallizing systems under various conditions with bench scale mixed suspension mixed product removal apparatus. The procedure is satisfactory for this particular size apparatus, where the degree of mixing allows for close approximation of an ideally mixed suspension. However, in large industrial apparatus the degree of mixing is far from ideal. Bulk concentrations can vary greatly throughout the reactor, and the residence time distribution can vary greatly from the ideal backmix reactor. In addition, particles may have differential settling velocities, and there may be a distinct difference in residence times for the two phases present. As a result of these nonidealities, growth rate and nucleation rate are usually modified to account for the degree of mixing, particle size, and heterogeneous effects.

An alternate approach for systems exhibiting kinetic behavior as Equations (1) and (2) would be to keep the kinetic relations, which are found to hold under ideally mixed conditions, but to employ different mixing models. One extreme in flow through a continuous reactor is to assume that the suspension acts as a macrofluid. Levenspiel (*2*) defines a macrofluid as a fluid which is made up of many small drops which are segregated or do not intermix. Thus it is assumed that these small drops act as small batch crystallizers, and for any given residence time distribution the averaged population density is given by

$$\overline{n(L)} = \int_0^\infty n\,(L, t)\, P\,(t)\, dt \tag{13}$$

GROWTH RATE

Consider now a crystallization system where the inlet liquor is cooled to the bulk temperature, immediately upon entering the crystallizer, or in the case of a salting out crystallizer, assume that the two streams are intimately mixed immediately upon entry to the crystallizer but macromixed thereafter. In either of these instances, let r_0 be the growth rate corresponding to the driving force represented by the difference in the input concentration and the equilibrium concentration. The following initial conditions are applicable to each of the tiny batch crystallizers which make up the macrofluid being processed:

$$\mu_0\,(0) = 0 \qquad \mu_2\,(0) = 0$$
$$\mu_1\,(0) = 0 \qquad r\,(0) = r_0$$

These boundary conditions are somewhat unrealistic, since in the case of continuous salting out crystallizers, the two streams usually enter the crystallizer separately.

Hence, if each were to act as a macrofluid, no crystallization would take place. In the case of a continuous cooling crystallizer, there would be some time delay until the tiny droplet reached the average temperature within the vessel. However, the authors feel that these limitations can be justifiably overlooked in order to obtain a rough estimate of this limiting extreme in flow through a continuous crystallizer.

In order to simplify the batch crystallization moment equations and mass balance, the new variable ψ and its transpose $t(\psi)$ are introduced:

$$\psi = \psi(t) = \int_0^t r(t')\, dt' \quad \text{and} \quad t = t(\psi) \tag{14}$$

By utilizing the new variable ψ to change the arguments in the pertinent function of the batch crystallization system, the following is defined:

$$\mu_i(t) = U_i(\psi) \qquad i = 0, 1, \ldots$$
$$r(t) = R(\psi).$$

With the aid of the new variable ψ, the following relationship is obtained for the batch crystallizer growth rate:

$$\frac{d^4 R}{d\psi^4} = -6 K_1 K_2 R^{i-1} \tag{15a}$$
$$R(0) = r_0$$
$$\frac{d^i R}{d\psi^i}(0) = 0 \qquad i = 1, 2, 3$$

from which the following relationship between t and ψ is found:

$$t = \int_0^{\psi} \frac{d\psi'}{R(\psi')} \tag{15b}$$

The cases when $i = 1$ and $i = 2$ will be considered. When $i = 1$, the batch crystallizer growth rate is given by

$$R(\psi) = r_0 - \frac{6 K_1 K_2}{4!} \psi^4 \tag{16a}$$

and from Equation (15*b*) ψ is related to t by

$$\frac{(\psi + \beta)}{(\beta - \psi)} e^{2 \tan^{-1} \psi/\beta} = e^{4 r_0 t/\beta} \tag{16b}$$

The parameter β is given by

$$\beta = \left(\frac{4 r_0}{K_1 K_2}\right)^{1/4} = \left(\frac{4 r_0}{k_n K_v \rho_c}\right)^{1/4} \tag{17}$$

When $i = 2$, the batch crystallizer growth rate is found to be

$$R(\psi) = r_0 \cos \phi\psi \cosh \phi\psi \tag{18a}$$

and ψ and t are related by

$$\int_0^{\psi} \frac{d\psi}{\cos \phi\psi \cosh \phi\psi} = r_0 t \tag{18b}$$

The parameter ϕ is given by

$$\phi = \left(\frac{3 K_1 K_2}{2}\right)^{1/4} = \left(\frac{3 k_n K_v \rho_c}{2 k_g}\right)^{1/4} \tag{19}$$

It is of interest to note the significance of the variable $\psi(t)$. If a crystal is produced at time zero by nucleation, then the size of this particular crystal at time t is simply $\psi(t)$. Hence, at the end of a batch crystallization the final population density function will only be defined on the crystal size interval $[0, \psi(\infty)]$ and will be zero elsewhere. This property, shown here only for the cases of $i = 1$ and $i = 2$, contrasts with the fact that the backmixed crystallization population density function at steady state is defined for all crystal sizes greater than or equal to zero. The maximum sized crystal which can be found at time infinity for $i = 1$ is

$$\max L = \left(\frac{4 r_0}{k_n K_v \rho_c}\right)^{1/4}$$

and for $i = 2$ is

$$\max L = \frac{\pi}{2} \left(\frac{2 k_g}{3 k_n K_v \rho_c}\right)^{1/4}$$

BATCH POPULATION DENSITY

With the initial conditions that all the moments are zero at time equal to zero, the general moment Equations (8) for a batch crystallization can be condensed by successive integration into the following form:

$$\mu_j(t) = K_1 j! \int_0^t r(\theta) \int_0^{\theta} r(\beta) \ldots r(\phi) \int_0^{\phi} r^i(\lambda)\, d\lambda\, d\phi \ldots d\rho\, d\theta$$

in which there are $j + 1$ integrations. Again, by utilizing the function $\psi(t)$

$$\psi(t) = \int_0^t r(t)\, dt$$

along with Dirichlet's formula for the inversion of multiple integrals, the above equation can be rearranged as

$$\mu_j(t) = \int_0^{\psi(t)} K_1 R^{i-1} [\psi(t) - L]\, L^j\, dL \tag{20}$$

Since $\psi(t)$ is the largest sized particle present in a batch crystallization at time t, $\mu_j(t)$ can also be written as

$$\mu_j(t) = \int_0^{\psi(t)} n(L, t)\, L^j\, dL \tag{21}$$

By comparison, the dynamic population density within a batch crystallizer is

$$n(L,t) = K_1 R^{i-1} [\psi(t) - L] \{1 - u[L - \psi(t)]\} \quad (22)$$

Equation (22) satisfies the original partial differential Equation (7) and will produce the correct dynamic moments, see Equations (20) and (21). Equation (22) also satisfies the condition of

$$n(0,t) = K_1 r^{i-1}(t)$$

In addition, the identical solution obtains by solving Equation (10) by the method of characteristics by using the above boundary condition.

MACROMIXED DISTRIBUTION, $i = 1$

From Equation (22), the batch population density function, for $i = 1$, is

$$n(L,t) = K_1 \{1 - u[L - \psi(t)]\} \quad (23)$$

where $\psi(t)$ is defined by Equation (14). The term $u[L - \psi(t)]$ is the unit step function delayed until L is equal to $\psi(t)$. At any time t, Equation (23) indicates that the batch crystallizer has a population density which is a straight line of value K_1 with slope zero that extends from $L = 0$ to $L = \psi(t)$.

To obtain the averaged population density function for the macrofluid crystallization taking place within a continuous crystallizer, it is necessary to define the residence time distribution for the crystallizer so that it can be applied in the averaging integral, Equation (13). Here, the residence time distribution will be taken as the ideal continuous stirred-tank residence time distribution:

$$P(t) = \frac{1}{\tau} e^{-t/\tau}$$

Upon application of this residence time distribution with the population density of Equation (23) in the averaging integral, the averaged population density obtained is

$$\overline{n(L)} = K_1 \left[\frac{(1 + L/\beta)}{(1 - L/\beta)} e^{\text{atm.}^{-1}(L/\beta)}\right]^{-\beta/4r_0\tau} \quad (24)$$

where L is on the interval $(0, \beta)$.

BACKMIXED CRYSTALLIZER, $i = 1$

The population density for the backmixed crystallization is

$$n(L) = K_1 e^{-L/r\tau} \quad (12a)$$

which is simply Equation (12) with the nuclei population density set equal to K_1. To compare this ideal backmix crystallizer distribution with Equation (24), a mass balance is taken around the backmix crystallizer:

$$6\rho_c K_V K_1 r^4 \tau^4 = C_0 - C$$

C_0 is the inlet bulk concentration. By using the factor $(\beta/4r_0\tau)$ in Equation (24), the above mass balance can be rearranged as

$$\left(\frac{4r_0\tau}{\beta}\right)^4 = \frac{4^3}{6}\left(\frac{r_0}{r}\right)^3\left(\frac{r_0}{r} - 1\right) \quad (25)$$

Figure 1 compares the backmix distribution, Equation (12), and macrofluid distribution, Equation (24), for values of r_0/r of 2 and 5 with the corresponding values for $(\beta/4r_0\tau)$ of 0.329 and 0.117, respectively. In this plot, the two choices of r_0/r may be regarded as two choices of residence time with the inlet concentration fixed. The comparison was done on this basis because of the complicated way that β is a function of the inlet bulk concentration. As would be expected, both the macromixed distribution and the backmixed distribution increase as the residence time is increased.

Graphical approximations of the zeroth moment and the first moment were made from Figure 1, and the following table was constructed to compare the total numbers produced by both the backmix crystallizer and the macromixed crystallizer, as well as to compare the numbers averaged size for both distributions, $\beta/4r_0\tau = 0.329$ and $\beta/4r_0\tau = 0.117$.

$\beta/4r_0\tau$	0.329		0.117	
	$\mu_0/K_1\beta$	$\bar{L}/\beta$	$\mu_0/K_1\beta$	$\bar{L}/\beta$
Backmix	0.380	0.380	0.427	0.427
Macromixed	0.5135	0.353	0.780	0.453

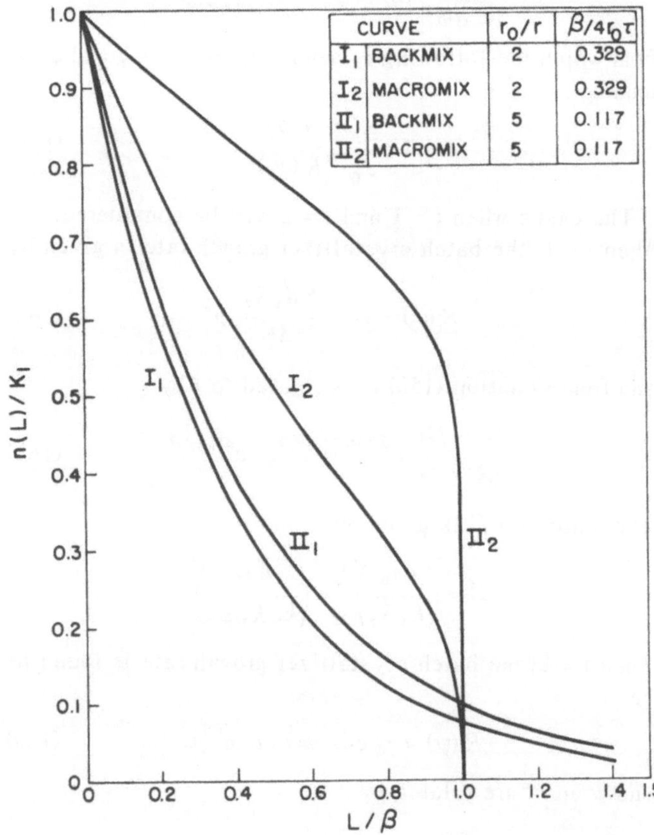

Fig. 1. Comparison of size distribution from a backmixed and macromixed crystallizer: $i = 1$.

By remembering that the constant $\beta/4\,r_0$ is held constant and that the residence time was varied, it can be seen that at both residence time levels the macromixed distribution has a higher total number of particles present. However, at the lower residence time level the backmixed numbers averaged particle size is greater than the macromixed numbers averaged particle size, but the reverse is true at the higher level of residence time.

If one knows that the solution being precipitated acts as a macrofluid, then for any given residence time distribution the averaged population density function can be found by applying the averaging integral Equation (13). It is of interest to find out what effect the residence time distribution has on the final averaged population density. To explore this behavior the following two residence time distributions are considered:

$$P_1(t) = \frac{1}{27.18}\, e^{-t/27.18}$$

and

$$P_2(t) = \frac{1}{12.82}\,(e^{-t/20} - e^{-t/7.18})$$

$P_1(t)$ is the standard residence time distribution for a single continuous ideally mixed vessel, and $P_2(t)$ is the residence time distribution for two continuous ideally mixed vessels in series. Both of these residence time distributions have the same total residence time, τ = 27.18. These types of models have been used to fit experimentally determined residence time distributions by Heng (*11*). He found that as the level of agitation is decreased, the residence time distribution for a single continuous stirred tank can be approximated by a series of ideally mixed tanks.

Figure 2 shows the crystal distribution under macromixed conditions which one might expect from residence time distributions $P_1(t)$ and $P_2(t)$ for a fixed value of $\beta/4\,r(0) = 2.34$. This plot indicates that the population density is increased over the entire size range applicable to the macromixed distribution. In Figure 2, the two comparable backmixed distributions are also presented. The comparable backmix distribution for $P_2(t)$ assumes that the tank with residence time equal to 20 leads the one of residence time equal to 7.18. The equations used in plotting Figure 2 are listed in Table 1.

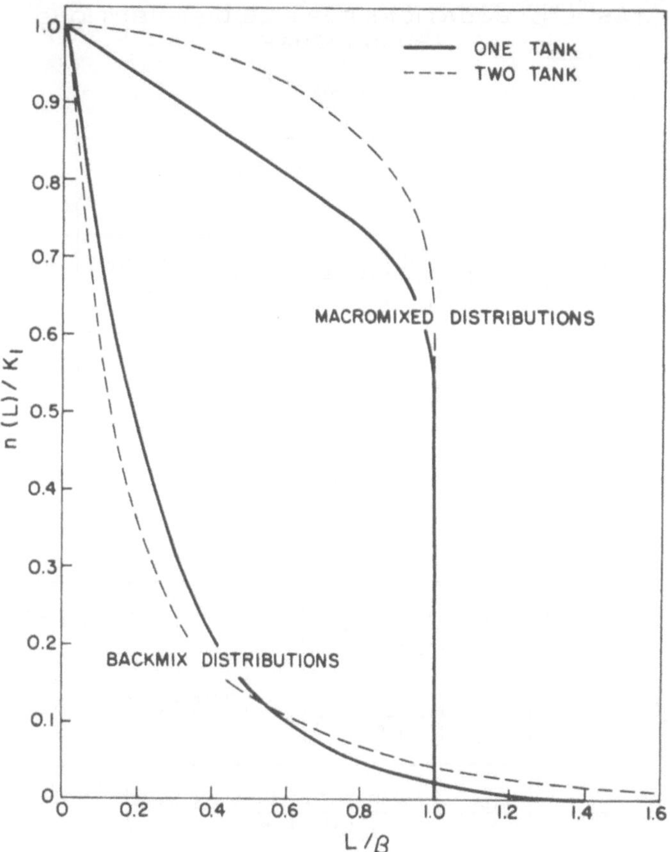

Fig. 2. Comparison of size distributions from a macromixed system with different residence time distributions: $i = 1$.

GENERAL COMMENTS ON MACROMIXING

A simplified method of obtaining the dynamic population density function for a constant volume, constant temperature, and well-mixed batch crystallization was developed for the case when the initial charge to the batch crystallizer is unseeded. The system kinetics are given by Equations (1) and (2). Investigation of the cases when the orders of the nucleation rate are 1 and 2 reveals that the resulting final batch population density function is defined on a finite size interval, $(0, L_{\max})$, and is zero elsewhere. This fact is not extended to cases where the order of the nucleation rate is greater than 2 in this paper, but it can be proved that this fact holds for all orders.

Utilization of the results for the case when the order of the nucleation rate is 1 yields the macromixed population density function for an ideal continuous mixed vessel. Figures 1 and 2 were constructed to show the behavior of a macromixed crystallizer as compared with an equivalent backmix crystallizer for various operating conditions. In general, these plots show that the macromixed population density function is only defined on a finite size interval, whereas the equivalent backmixed population density function is defined for all crystal sizes greater than or equal to zero. This fact can also be extended to all orders of the nucleation rate, since it can be shown that the final batch population density function is always defined on a finite interval. In addition to this property, it can be shown that the ratio of the mass averaged size of macromixed crystallization to that of the backmixed crystallizer for the case when the order of the nucleation rate is 1 obeys the inequality

TABLE 1. EQUATIONS FOR SIZE DISTRIBUTIONS USED IN FIGURE 2

1 TANK

Backmix:	$\frac{n(L)}{K_1} = e^{-4(r_o/r)(\beta/4r_o\tau)L/\beta}$
Macromix:	$\frac{n(L)}{K_1} = \left[\frac{(1 + L/\beta)}{(1 - L/\beta)} e^{2 \tan^{-1} L/\beta}\right]^{-\beta/4r_o\tau}$
Parameters:	$\frac{\beta}{4r_o} = 2.34, \frac{r_o}{r} = 11.25, \frac{\beta}{4r_o\tau} = 0.0862$

2 TANK

Backmix:	$\frac{n(L)}{K_1} = \left[\frac{r_1\tau_1/r_2\tau_2 - 2}{r_1\tau_1/r_2\tau_2 - 1}\right] e^{-4(r_o/r_1)(r_1/r_2)(\beta/4r_o\tau_2)L/\beta} + \left[\frac{1}{r_1\tau_1/r_2\tau_2 - 1}\right] e^{-4(r_o/r)(\beta/4r_o\tau_1)L/\beta}$
Macromix:	$\frac{n(L)}{K_1} = (\frac{\tau_1}{\tau_1 - \tau_2}) \left[\frac{(1 + L/\beta)}{(1 - L/\beta)} e^{2 \tan^{-1} L/\beta}\right]^{-\beta/4r_o\tau_1} - (\frac{\tau_2}{\tau_1 - \tau_2}) \left[\frac{(1 + L/\beta)}{(1 - L/\beta)} e^{2 \tan^{-1} L/\beta}\right]^{-\beta/4r_o\tau_2}$
Parameters:	$\frac{r_o}{r_1} = 5, \frac{r_1}{r_2} = 1.302, \frac{\beta}{4r_o} = 2.34$

$$\frac{L_{Mm}}{L_{Bm}} < 1.16 \left(\frac{r_0}{r_0 - r}\right)$$

This inequality is rather inconclusive but does indicate that one might obtain a higher mass averaged crystal size in a macromixed crystallization than in a backmixed crystallization when the order of the nucleation rate is 1.

Again, a macrofluid represents an extreme in fluid conditions. A more likely situation would be that the fluid as it just enters a continuous crystallizer is in the macrofluid region of flow and then disperses into a backmix region.

Consider one additional model which incorporates delayed mixing of the macrofluid droplets. Nonideal flow within a continuous stirred crystallizer might have the following properties:

1. The fluid enters the reactor as a macrofluid.
2. The small droplets have a mean life expectancy as a macrofluid of τ_L.
3. The total fluid within the reactor is made up of a macrofluid portion and an ideal mixed portion. The droplets can be thought of as distributed within the reactor by the ideally mixed phase much as drops of oil in water.

The additional probability of leaving the macrofluid flow category along with the existing probability of leaving the reactor by the outlet stream results in an age distribution of the macrofluid phase of

$$P(t) = \frac{1}{\tau^*} e^{-t/\tau^*} \tag{26}$$

where $\tau^* = \tau\tau_L/(\tau + \tau_L)$.

Equation (26) is also the age distribution of the feed for the backmix section. Therefore, this mixed model can be diagramed as shown in Figure 3. The volumetric fraction that leaves the reactor as a macrofluid f_v is equal to τ^*/τ.

The delayed mixing concept has some satisfying aspects, and work continues on its development. Size distributions from the more realistic case where dispersion and recoalescence take place would be between the extreme shown here for completely backmixed and completely macromixed. This case has not been solved. This development on macromixing illustrates perhaps an unrealistic limiting case, but it does demonstrate how mixing conditions such as might exist in a highly viscous system effect size distribution.

MIXED MODELS APPROACH

Another method sometimes employed to compensate for nonideal flow conditions is the mixed models approach. Essentially, the mixed models approach assumes that flow within a continuous reactor can be broken up into various interconnected regions where perhaps plug flow or possibly backmix flow might closely approximate the true flow characteristics.

Consider the idealized system shown in Figure 4.

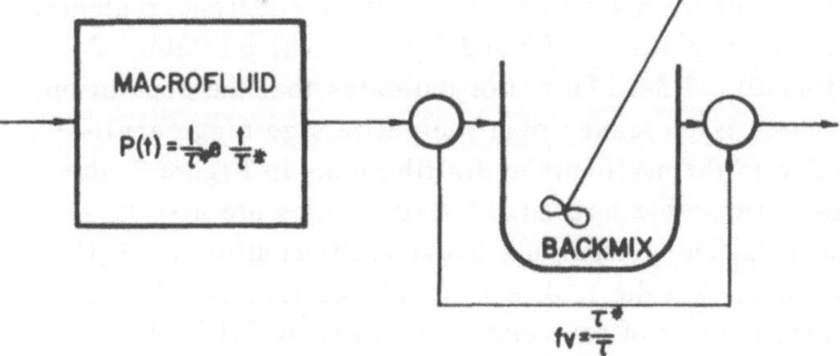

Fig. 3. Delayed mixing crystallizer.

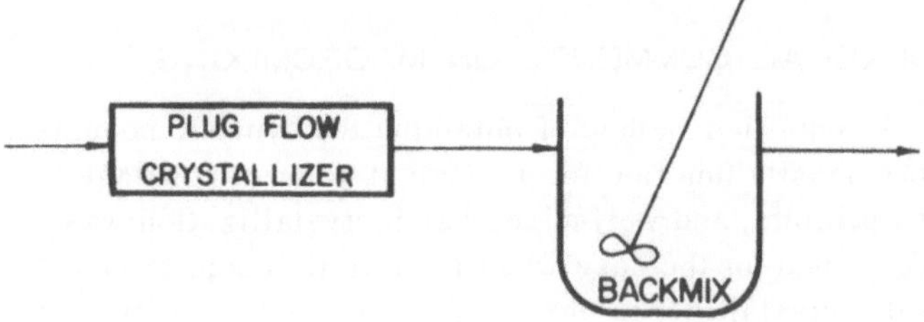

Fig. 4. Idealized plug flow-backmix crystallizer.

This system consists of a plug flow section followed by a well-backmixed section. In order to simplify this treatment, solvent, solute, and heat balances are neglected. Only the population densities leaving the ideal backmix crystallizer are determined. It is also assumed that the volume changes due to crystallization are negligible ($K_V\mu_3 << 1$) and that the crystal density is much larger than the bulk concentration in the same units ($\rho_c >> C$). Finally, only the case when the order of the nucleation rate 2 is considered.

The plug flow particle balance, Equation (5), is

$$v_x \frac{\partial n}{\partial x} + \frac{\partial (nr)}{\partial L} = 0$$

As noted in the discussion of moment balances, the moment equations and the particle balance equation for a plug flow reactor are identical to those of an ideal batch reactor if x/v_x is substituted for time. Therefore, the growth rate at any distance x along the plug flow crystallizer is given by

$$r = r_0 \cos [\phi\psi(x)] \cosh [\phi\psi(x)]$$

where $\psi(x)$ is

$$\psi(x) = \int_0^x r dx/v_x$$

The population density at any axial distance along the plug flow crystallizer is

$$n(x, L) = K_1 r_0 \cos [\phi\psi(x) - \phi L] \cosh [\phi\psi(x) - \phi L]$$

The above population density evaluated at the plug flow exit serves as a feed to the backmix crystallizer. The particle balance about the backmix crystallizer is

$$r \frac{dn}{dL} = \frac{1}{\tau} [n(x, L) - n]$$

The resulting population density leaving the backmix crystallizer can be broken up into two parts. The first part is on the particle size range $[0, \psi(x)]$ and is

$$n(L) = K_1 r e^{-L/r\tau} + \frac{K_1 r_0}{2\phi r\tau} \{\sin \xi_1 \cos (\phi\psi + \xi_1 - \phi L) e^{\phi(\psi - L)} + \sin \xi_2 \cos (\phi\psi + \xi_2 - \phi L) e^{-\phi(\psi - L)} - [\sin \xi_1 \cos (\phi\psi + \xi_1) e^{\phi\psi} + \sin \xi_2 \cos (\phi\psi + \xi_2) e^{-\phi\psi}] e^{-L/r\tau}\} \quad (27)$$

and the second part is for $L > \psi(x)$ and is

$$n(L) = \left\{K_1 r + \frac{K_1 r_0}{2\phi r\tau} [\sin \xi_1 \cos \xi_1 e^{\psi/R\tau} + \sin \xi_2 \cos \xi_2 e^{\psi/R\tau} - [\sin \xi_1 \cos (\phi\psi + \xi_1) e^{\phi\psi} + \sin \xi_2 \cos (\phi\psi + \xi_2) e^{-\phi\psi}]]\right\} e^{-L/r\tau} \quad (28)$$

The constants ξ_1 and ξ_2 are given by the following relations:

$$\xi_1 = \tan^{-1} \left[\frac{\phi r\tau}{1 - \phi r\tau}\right] \qquad \xi_2 = \tan^{-1} \left[\frac{\phi r\tau}{1 + \phi r\tau}\right]$$

There are three dominant constants which appear in the above equations ($\phi\psi$, $\phi r\tau$, r_0/r). By means of a mass balance about the backmix crystallizer, these three constants are related by the following equation:

$$4(r\tau\phi)^4 + 1 = \frac{r^\circ}{r} [\cos \phi\psi \cosh \phi\psi - (r\tau\phi)(\sin \phi\psi \cosh \phi\psi - \cos \phi\psi \sinh \phi\psi) - 2(r\tau\phi)^2 \sin \phi\psi \sinh \phi\psi - 2(r\tau\phi)^3(\cos \phi\psi \sinh \phi\psi + \sin \phi\psi \cosh \phi\psi)] \quad (29a)$$

Finally, if the total time T_{TOT} and the initial growth rate r° are held constant, one obtains the additional relationship

$$\left(\frac{r^\circ}{r}\right)(r\phi\tau_{Bm}) + r^\circ\phi\tau_{PF} = T_{TOT}\, r^\circ\phi \quad (29b)$$

The graphs in Figures 5 and 6 were constructed by using Equations (27), (28), and (29a, b). The specific value of $T_{TOT}\, r^\circ\phi$ in Equation (29b) was taken as 4.1987. Figure 5 shows different population density functions which arise from various values of τ_{Bm}/T_{TOT}. For instance, the graph associated with τ_{Bm}/T_{TOT}

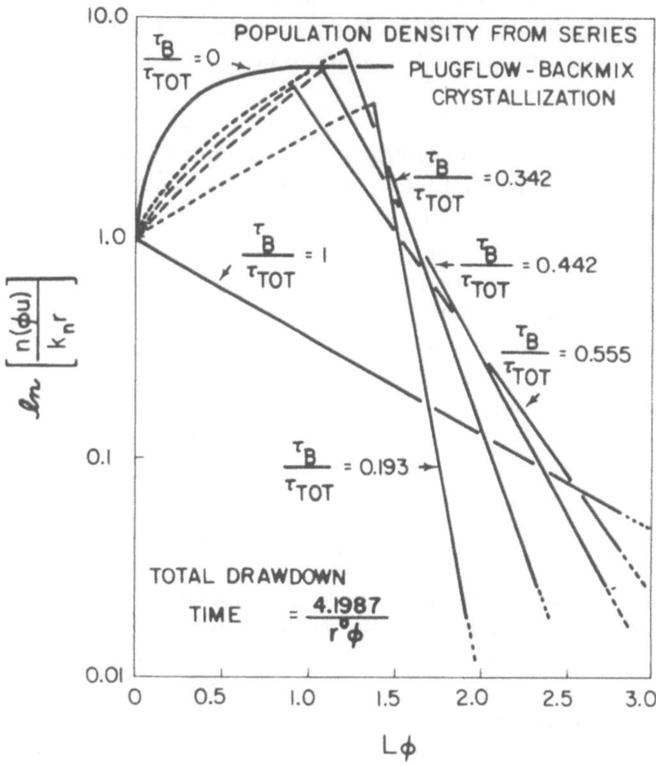

Fig. 5. Population density from a series backmix-plug flow crystallizer.

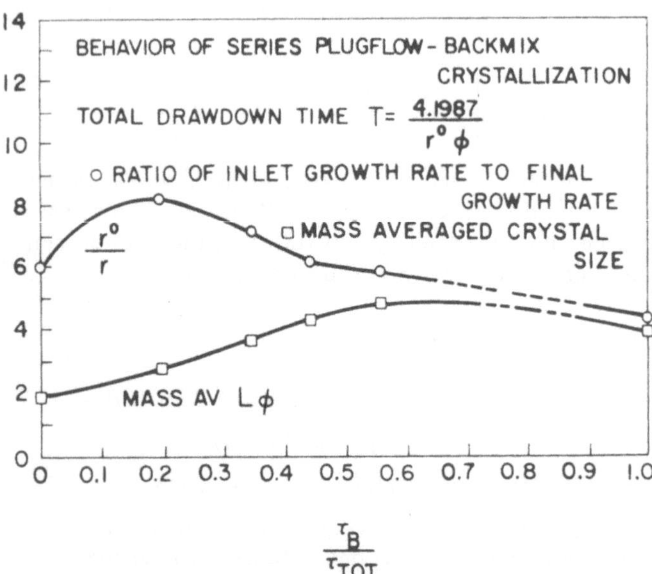

Fig. 6. Behavior of a series backmix-plug flow crystallizer.

equal to 1 corresponds to a single backmix crystallizer, and the graph associated with τ_{Bm}/T_{TOT} equal to zero is associated with a single plug flow crystallizer. Note that if flow is characterized by a series plug flow-backmix model, then extrapolation of the logarithmic portion of these graphs back to crystal size zero will give values of an apparent nuclei population density n_A° which are much higher than the actual nuclei population density n_0°. In this particular example, if τ_{Bm}/T_{TOT} is greater than 0.95, that is, if the backmix portion occupies more than 95% of the total volume of the system, then the ratio n_A°/n_0° is bounded by

$$1 \leq n_A^\circ/n_0^\circ < 2.35$$

This fact can lead to errors in continuous mixed suspension mixed product removal experiments as outlined by Larson and Randolph (9) if flow within the experimental vessel is actually a series plug flow-backmix system.

Figure 6 illustrates the behavior of the series plug flow-backmix crystallizers in terms of the parameters r°/r and $L_m\phi$ at various values of τ_{Bm}/T_{TOT}. The ratio r°/r is actually the ratio of the inlet supersaturation to the outlet supersaturation and hence is a measure of the mass of crystals produced. Note that in this particular example the highest mass conversion to crystals is obtained when τ_{Bm}/T_{TOT} is approximately equal to 0.18 or in a system in which the backmix portion occupies 18% of the total volume. The parameter ϕL_m is the dimensionless mass averaged crystal size, and in this particular case the mass averaged crystal size reaches a maximum at a value of τ_{Bm}/T_{TOT} of about 0.6 or in a system in which the backmix portion represents 60% of the total volume.

The mixed model presented here has some similarity to forced circulation crystallizers which have regions of plug flow and regions of mixing. The relevance of this model to these cases will depend on the relationship of the circulation time to the residence time. High circulation rates should give more nearly fully backmixed conditions.

DISCUSSION

The models presented in this paper are not now directly applicable to commercial systems. They do, however, illustrate the limiting distribution for two classical concepts of poor mixing. One could conclude that real systems should give results between the backmixed model and the completely macromixed model or the combined plug flow-backmix model. It is reasonable, however, that size distributions outside these limits could be obtained if, for instance, the mixing condition for the solid phase differed from that of the liquid phase. This is undoubtedly a very real problem in large systems. Substantial attrition, of course, will also cause the size distribution to differ from that predicted by these equations.

These models should be relevant in considering certain problems in crystallization. For example, if it is assumed that regions of high supersaturation exist near the feed point, the plug flow-backmix model should be useful, as well as the idea of the two tank residence time distribution function. In the latter case one would logically assume a smaller tank as the first and the larger tank as the second. This is just the reverse of that considered here.

As mentioned before, the mixed model should be useful in explaining forced circulation systems, and the completely macromixed should be useful in studying viscous systems.

Much work remains to be done on this very complicated problem. Expansion on each of these ideas is necessary to fully come to grips with the very real problems found by those designing and operating industrial systems.

ACKNOWLEDGMENT

The authors wish to acknowledge the National Aeronautics and Space Administration, the National Science Foundation by Grant GK 1328, and the Iowa State University Engineering Research Institute for their support of this work.

NOTATION

C = bulk solution concentration, mass of crystalline dissolved/unit volume

C_{eq} = equilibrium concentration, mass of crystalline dissolved/unit volume

C_0 = inlet bulk concentration

dL/dt = growth rate
dN°/dt = nucleation rate
i = kinetic order of nucleation rate
k_g = proportionality constant for growth rate
k_n = proportionality constant for nucleation rate
K_v = volume shape factor
L = particle size, linear
n = population density function, numbers/unit size/unit volume
n° = population density of nuclei ($L = 0$)
n_i = population density function, input to backmix crystallizer
$n(L, t)$ = batch reactor population density at any time
$\bar{n}$ = averaged macromixed population density function
$\overline{n(L)}$ = overall or average population density
$P(t)$ = residence time distribution
r = growth rate
r_0 = growth rate evaluated at C_0
r_0 = growth rate at entrance to plug flow crystallizer
t = reactor time
v_x = axial plug flow velocity
$\bar{v}$ = fluid velocity
x = axial length of plug flow crystallizer
$\bar{x}$ = coordinate space within reactor

GREEK LETTERS

$\psi(t)$ = integral of batch crystallization growth rate evaluated between $t = 0$ and t
$\psi(x)$ = plug flow equivalent to $\psi(t)$, $t = x/v_x$
ρ_c = crystal density
τ = residence time of reactor
τ_i, τ_0 = residence times of the vessel based on inlet and outlet flow velocities, respectively
τ_L = mean life of the macrofluid

LITERATURE CITED

1. Powers, H. E. C., *Ind. Chem.*, **39**, 351–355 (1963).
2. Levenspiel, O., "Chemical Reaction Engineering," Wiley, New York (1964).
3. Curl, R. L., *AIChE J.*, **9** (1963).
4. Shinnar, R., S. Katz, and J. J. Evangelista, *ibid.*, to be published.
5. Neilsen, A. E., "Kinetics of Precipitation," Pergamon Press, New York (1964).
6. Canning, T. F., and A. D. Randolph, *AIChE J.*, **13**, 5–10 (1967).
7. Carey, N. W., and J. Estrin, *Ind. Eng. Chem. Fundamentals*, **6**, 13–20 (1967).
8. Timm, D. C., PhD thesis, Iowa State Univ., Ames (1967).
9. Randolph, A. D., and M. A. Larson, *AIChE J.*, **8**, 639–645 (1962).
10. Murray, O. A., and M. A. Larson, *ibid.*, **11**, 728–733 (1965).
11. Heng, O. A., PhD thesis, Iowa State Univ., Ames (1966).

AN EXPERIMENTAL STUDY OF CRYSTAL SIZE DISTRIBUTION IN A CONTINUOUS, BACKMIXED, GAS-PHASE REACTOR

Paul D. Stone
and
Alan D. Randolph

An experimental study was made of the crystal size distribution produced in the gas-phase crystallization of ammonium chloride in a continuous, single-stage, backmixed reactor. Linear growth rates were investigated as a function of the holding time, and an overall crystallization mechanism for this system was deduced by means of electron microscopy data and experimental nucleation-growth rate data. Results from this work indicate that nucleation and growth rate kinetics can be studied in such gas-phase systems by measuring crystal size distributions and gas-phase compositions under a variety of operating conditions.

One of the most important aspects of a process producing a solid-phase product is the particle size distribution obtained in the process. Although much has been done on liquid-phase crystallization, and on single crystal growth from the gaseous phase, little work has been reported in the area of particle size distributions from a continuous, gas-phase process. Discussed in this paper will be the efforts of the authors in designing and utilizing a single-stage, backmixed reactor to study crystal size distribution (CSD) and growth-nucleation rate kinetics in a continuous, gas-phase crystallization system. A specific motivation for such gas-phase studies is the evaluation of the isothermal backmixed reactor as a kinetic tool in the study of other combustion systems which produce particulate products.

The crystallization system in this research effort consisted of gaseous hydrochloric acid and gaseous ammonia reacting in a nitrogen carrier gas to produce ammonium chloride. The sampling techniques, counting procedures, and overall experimental design philosophy were oriented towards electron microscopy fundamentals, since the electron microscope was a basic necessity for studying the ammonium chloride crystals produced. The ammonium chloride system was chosen for this study mainly because hydrochloric acid and ammonia react readily at room temperature and release a relatively small heat of formation when compared with standard hydrocarbon combustion systems. Therefore, low-temperature isothermal conditions could be easily maintained. Also, ammonium chloride deposits could be easily rinsed out of the reactor with water after each experimental run.

The emphasis of this paper is on the experimental techniques and approach necessary in such a study.

OBJECTIVES

The overall objective of this experimental program was to study CSD formed in such a gas-phase system under a variety of conditions and to evaluate the feasibility of obtaining quantitative reaction kinetics from such data. An isothermal backmixed reactor might turn out to be a valuable tool in studying combustion processes producing particulate products. An additional objective was the development of particle sampling, counting, and crystallographic techniques for the study of crystal particles in the 0.1 to 10 μ size range.

The primary task was to observe if theoretically predicted crystal size distributions actually describe the real situation. Specifically, population balance derived

University of Florida, Gainesville, Florida. Alan D. Randolph is at the University of Arizona, Tucson, Arizona. Paul D. Stone is with Dow Chemical Company, Midland, Michigan.

size distributions were to be compared with those produced in the gas-phase crystallization of ammonium chloride. It was also desired to determine the experimental nucleation rates and growth rates of this gas-phase process as a function of the reactants' partial pressures.

Thus

$$r = f(p_1, p_2)$$

and

$$dN^\circ/dt = g(p_1, p_2)$$

The partial pressures are a measure of the molecular driving forces which bring about crystal nucleation and growth.

Although the growth rates as a function of holding time and nuclei population density as a function of the growth rate were investigated, the objective of correlating nucleation-growth rate kinetics with partial pressures was not accomplished in this present study because of a lack of time. However, important experimental procedures and techniques were established which show the feasibility of investigating such gas-phase systems in this manner. As was concluded by Randolph (6), much more work is needed on the very difficult problem of determining nucleation and growth rate kinetics in the actual environment of a mixed crystal suspension. Such kinetic data is necessary in order that one may specify and design a system to produce the desired CSD. The starting point for this work is believed to be the population balance approach to CSD data.

POPULATION BALANCE

The theoretical basis for this study was the population balance. In the engineering sciences, the concepts of mass, energy, and momentum balances are well known and accepted. Randolph has discussed the basic concept of a population balance for countable entities, a concept which has considerable practical value in crystallization systems (5). Although there are several ways to express the size distribution of particles along a size axis, the population density (number distribution) is the most fundamental way. Thus, the population density or number distribution function $\bar{n}(L)$ is defined such that the number of crystals present in a small size range ΔL of a unit volume of suspension is given by $\Delta N = \bar{n}\Delta L$. The units of $\bar{n}$ in this paper are the number of crystals per cubic centimeter of suspension per micron of length.

Randolph and Larson (4) have derived a generalized population balance for an arbitrary crystal suspension which can be reduced to the simple case considered here.

The general equation is given as

$$\int \left[\frac{\partial \bar{n}}{\partial t} + \frac{\partial}{\partial L}\bar{n}\left(\frac{\partial L}{\partial t}\right)\right] dV + \bar{n}_s \frac{dV}{dt} = \frac{R_i \bar{n}_i}{\rho_i} - \frac{R_0 \bar{n}_0}{\rho_0} \tag{1}$$

which, for the single-stage, mixed suspension, mixed product removal, gas-phase crystallizer studied in this research, reduces to

$$\frac{\partial n}{\partial t} = -r\frac{\partial n}{\partial L} - n/T \tag{2}$$

where $n = \bar{n}V$ = number of crystals per micron in the vessel.

The following restrictions were applied in deriving Equation (2):

1. $\partial L/\partial t = r \neq f(L)$, assuming McCabe's ΔL law is applicable.
2. $\bar{n}_i = 0$, gaseous feeds only.
3. $\bar{n}_0 = \bar{n}$, because of mixed product removal.
4. $dV/dt = 0$, for constant suspension volume.
5. $T^{-1} = R^\circ/\rho_0 V$, the reciprocal of holding time, or drawdown time.

If steady state conditions are maintained, then $\partial n/\partial t = 0$, and one obtains the following size distribution solution to Equation (2):

$$\ln n = -L/rT + \ln n^\circ \tag{3}$$

or

$$n = n^\circ \exp(-L/rT) \tag{4}$$

Therefore, Equations (3) and (4) represent the steady state solution for the CSD from a mixed suspension, mixed product removal (MSMPR) crystallizer. The graphic form of the solution is shown in Figure 1. When

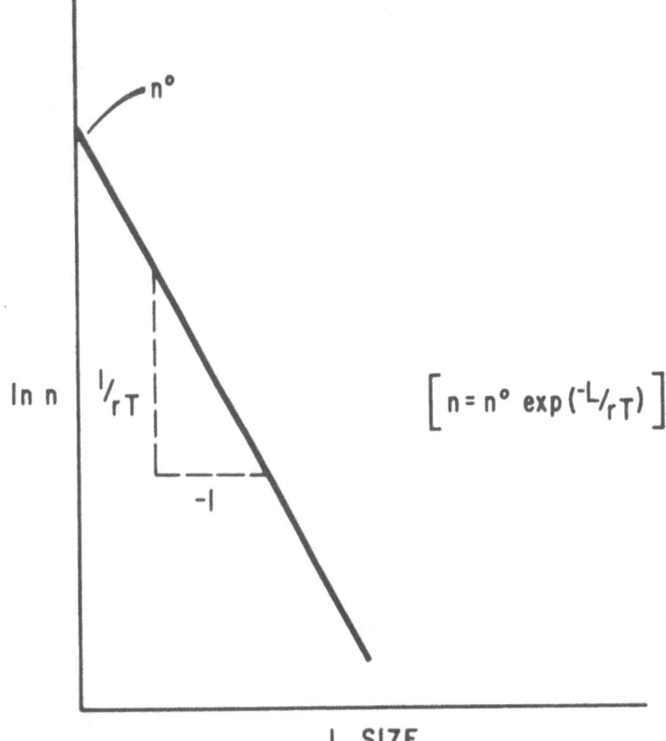

Fig. 1. Steady state solution to population balance for CSD from MSMPR crystallizer.

one knows the reactor volume and volume flow rates, then the average holding time T and the crystal growth rate r can easily be calculated.

Population density values of n in this paper were calculated from the following relation:

$n = \Delta N/\Delta L$ = total particles counted in size range divided by the size range

From the above definition, it follows that the total number of particles can be obtained by integration:

$$dN = ndL \tag{5}$$

$$N_T = \int_0^{\infty} ndL \tag{6}$$

The total surface area (square centimeters) of crystals per unit volume (cubic centimeters) of suspension can be expressed as

$$\overline{A} = k_A \int_0^{\infty} \overline{n}L^2 dL \tag{7}$$

Similarly, the volume of crystals per unit volume of suspension can be expressed as

$$\overline{V} = k_V \int_0^{\infty} \overline{n}L^3 dL \tag{8}$$

At steady state, there is a constant mass of crystals in the system suspension where

$$M_T = \frac{\rho_c k_V}{m} \int_0^{\infty} \overline{n}L^3 dL \tag{9}$$

The above properties can be reduced to the general size distribution moment form of

$$f_j = \int_0^{\infty} nL^j dL \tag{10}$$

With the steady state solution in Equation (4) and the above expression for M_T, one obtains the following:

$$M_T = \frac{\rho_c k_V}{m} (rT)^4 \overline{n}^{\circ} \tag{11}$$

The solids concentration M_T as given above occurs in a mass balance for the system. As M_T is obviously limited by the concentration of reactants fed to the system, Equation (11) then forms a constraint on the growth rate obtainable at a given nucleation rate (6).

NUCLEATION AND GROWTH

Gas-phase nucleation is the phenomenon of nuclei generation of a solid form capable of growing within the gaseous phase. In this research, the nuclei are formed as a direct result of the chemical reaction between the gaseous ammonia and hydrochloric acid:

$$NH_3(g) + HCl(g) = NH_4Cl(s) + \Delta H_f$$

$$\Delta H_f = 42.1 \text{ kcal./g.-mole (at 18°C. and 1 atm.)}$$

The critical nucleation determining embryo becomes a nucleus that differs from an equal number of other ammonium chloride molecules suspended throughout the gaseous phase by possessing a characteristic surface energy (Gibbs free energy) necessary for a stable, new growth center. It has been observed that mechanical mixing, reactor walls, other crystals, impurity particles, and many other conditions besides the reactants' partial pressures can have important effects on nucleation.

Nucleation and growth take place simultaneously. As other molecules attach themselves to the nucleus with a degree of regularity, the crystal grows in size. There are many growth mechanisms and types of irregularities which may occur in crystallization. In this work, both solid and dendritic types of crystals were grown in the gaseous phase. Basically, ammonium chloride is a colorless substance with a cubic structure and a density of 1.527 g./cc.

The dendritic form of crystallization in which dendrites or treelike formations are developed is one of the most common modes which crystals take in a natural environment (*2*). There are many possible explanations for dendritic growth which include preferred growth directions, surface energies, speed of disposition, and gas-phase properties which limit heat and mass transfer. Many substances such as ammonium chloride show a natural preference for growing in dendrite structures. However, a dendritic crystal which is filled in or built up theoretically can still become a good solid crystal. Dendritic growth is especially prevalent for the gas-phase crystallization of ammonium chloride as revealed in this study. Actually, dendritic structures exist in endless variations, all of which are grouped under the name of *dendrites*.

As was indicated in the preceding section, the growth rates determined in this research were assumed to follow McCabe's ΔL law. This law, formulated after many observations by McCabe, states that the crystal growth rate is independent of the crystal size for crystals submitted to the same external environment. This means that in a mixed crystal suspension, each crystal will grow the same length in a given time period when subjected to the same environment. However, this law is only an empirical statement of fact which may or may not be true in any particular system.

One of the objectives was to experimentally study the kinetics of growth rates and nucleation rates as a function of the partial pressures. Although very little has been published on data of this type for either gas-phase or liquid-phase crystallization, most liquid phase data has appeared to agree with the following simple power model form (*4, 6*): let

$$r \equiv dL/dt \tag{12}$$

Then, nucleation rate is given in terms of r by the power model relationship

$$dN^\circ/dt = k_N r^\alpha \equiv r\bar{n}^\circ \quad (13)$$

or

$$\bar{n}^\circ = k_N r^{\alpha-1} \quad (14)$$

It would be valuable to know how well such a power model form would apply to the analogous phenomena of gas-phase crystallization. This was included in the authors' primary objectives.

EXPERIMENTAL

As a result of preliminary experimental studies, the design of a spherical glass reactor with tangentially located nozzles for the reactants was chosen. This design satisfied the conditions of a continuous, mixed suspension, gas-phase crystallizer and still remained a flexible, simple system. The shape of the spherical reactor was such as to accomplish fairly complete backmixing by means of the positioning of the gas inlet nozzles. The configuration to be discussed is specifically for ammonium chloride formed from the gaseous phase.

DETAILS

The reactor consisted of a one liter, round bottom, Pyrex, long-necked flask. Five holes of approximately $^5/_{32}$-in. diameter were cut into the flask with an oxygen-gas torch: two holes for the nitrogen carrier gas, two holes for the reactants, and one hole offset for the thermometer and gas sampler. The holes were reinforced with a rubbery, silicone sealant in order to provide a flexible coupling seal for the glass nozzles. Four identical glass nozzles were used: one each for the ammonia and hydrochloric acid, and two on opposite sides for the nitrogen. A schematic of the reactor is shown in Figure 2. The actual reactor and apparatus arrangement are shown in Figure 3. Tygon tubing of $^1/_4$ in. I.D. was used throughout the apparatus, since it is compatible with both ammonia and hydrochloric acid. Flow rates were measured with standard calibrated flowmeters. All temperature measurements were accomplished with a standard thermometer inserted in the accessory hole with the mercury tip flush with the reactor wall such that crystal deposits would not easily accumulate. The reagents consisted of 99.99% ammonia, 99.0% hydrochloric acid, and 99.8% dry nitrogen.

a.) FLASK - SIDE VIEW

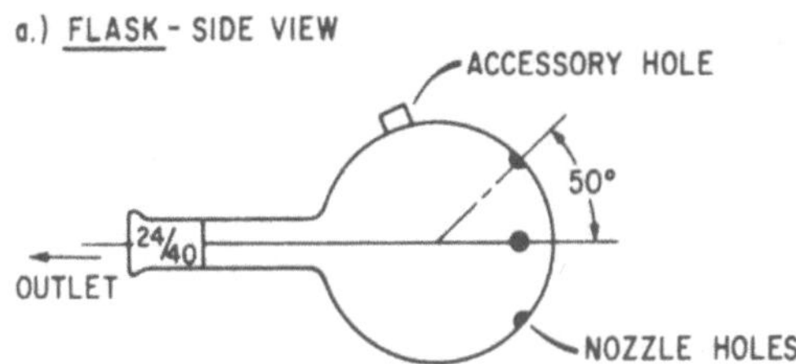

b.) FLASK - REAR VIEW SHOWING NOZZLE CONFIGURATION IN SCHEMATIC

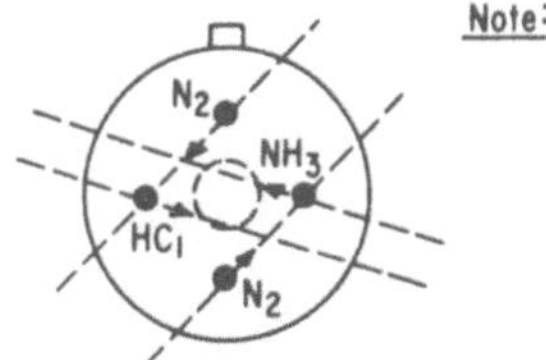

Note: ARROWS POINT ALONG PATH OF THE GAS NOZZLE JET. THE NOZZLE OUTLETS INSIDE THE REACTOR WERE APPROXIMATELY PARALLEL TO REACTOR WALL.

c.) NOZZLE - DRAWN FROM STANDARD LABORATORY GLASS TUBING

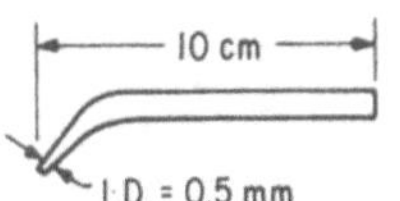

Fig. 2. Reactor.

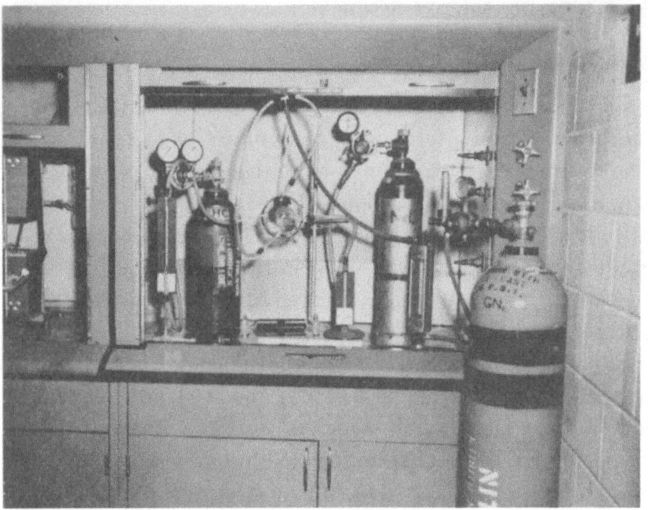

Fig. 3. Experimental laboratory reactor.

The reactor and nozzles were rinsed with water and acetone and then thoroughly dried before each run. The nozzles were always returned to the same arrangement as shown in Figures 2 and 3. The reactor was purged with dry nitrogen for 2 to 3 min. each run before reactants were admitted. After the purging, first the ammonia and then the hydrochloric acid were adjusted to the desired flow rates. Immediately after admitting the reactants, one could see the actual swirling and backmixing action inside the reactor. The reactor contents soon became a homogeneous, milky mixture with the crystal suspension gently flowing from the reactor outlet. Dependent upon how long the reactor operated, there was a slight accumulation of crystals clinging to some sections of the reactor walls; it is estimated that approximately one-half of the reactor's interior surface always remained free of crystal deposits.

Flowmeters were read as accurately as possible. They were adjusted periodically, if necessary, to maintain the flow rates at the desired values. After a manual adjustment of the flow, a time of approximately ten times the average holding time was allowed in order to achieve steady state. Also, crystal samples were not taken until the reactor was at steady state according to this empirical rule of thumb. Samples were taken by holding the specially prepared electron microscopy grid in the reactor outlet stream for 15 sec. This sampling time was determined to be optimum for obtaining a suitable density of ammonium chloride crystals on the sampling grid. The grid was held with fine tipped microscopy tweezers. A backup sample was taken for each run. Preliminary work was accomplished which indicated that no substantial crystal growth was taking place on the grid during the 15 sec. sampling period.

Temperatures determined from the standard thermometer were reasonably accurate since, it was found in preliminary tests that the thermometer did reach thermal equilibrium with the reactor contents.

The assembled apparatus was located inside a fume hood for convenience and safety.

ELECTRON MICROSCOPY

SIZING AND COUNTING

Much exploratory work was done in developing satisfactory electron microscopy procedures and techniques. These techniques will only be outlined in this paper since a full discussion would be very lengthy and detailed. A JEM-6A (Japan Electron Optics Laboratory) electron microscope was used in this effort.

A major problem was encountered with ammonium chloride crystals rapidly subliming under the electron beam of the microscope. The sublimation temperature of ammonium chloride is 340°C., and it is not uncommon for electron microscopy samples to reach 300 deg. and higher under an intense beam. Although many approaches were tried, this problem was overcome mainly by operator technique. The electron beam intensity was maintained as low as possible until a very thin layer (possibly 50 Å in thickness) of vacuum contamination had deposited on the sample crystals. Since the image was not significantly obscured, this contamination film was very useful in hindering sublimation when the beam intensity was finally increased for viewing and photographing.

The basic sample support substrates were prepared by coating glass microscope slides with formvar (a 2% solution of polyvinyl formal resins). This mounted film was then scored into squares, floated off onto water in a casting dish, and transferred to 200-mesh nickel electron microscopy screens by using the peg and ring technique. After the formvar film had dried on the screens, each covered screen was also coated with a very thin film of silicon monoxide by using a vacuum evaporator to increase the total substrate stability.

All electron microscopy data was recorded on 3¼ in. × 4 in. photographic plates of medium contrast. Although the interpretation of electron microscopy micrographs does involve special considerations, this area will not be discussed here.

Counting and sizing of the ammonium chloride crystals was done by manual and visual methods on the photographic plates. Usually four to six plates were exposed for each subrun, with only two plates being counted and sized for each subrun. Approximately two hundred crystals were usually visible on each plate. The crystal dimension that was measured was not a problem, since over 90% of the crystals studied were spherical. Most of the electron micrographs were taken at magnifications

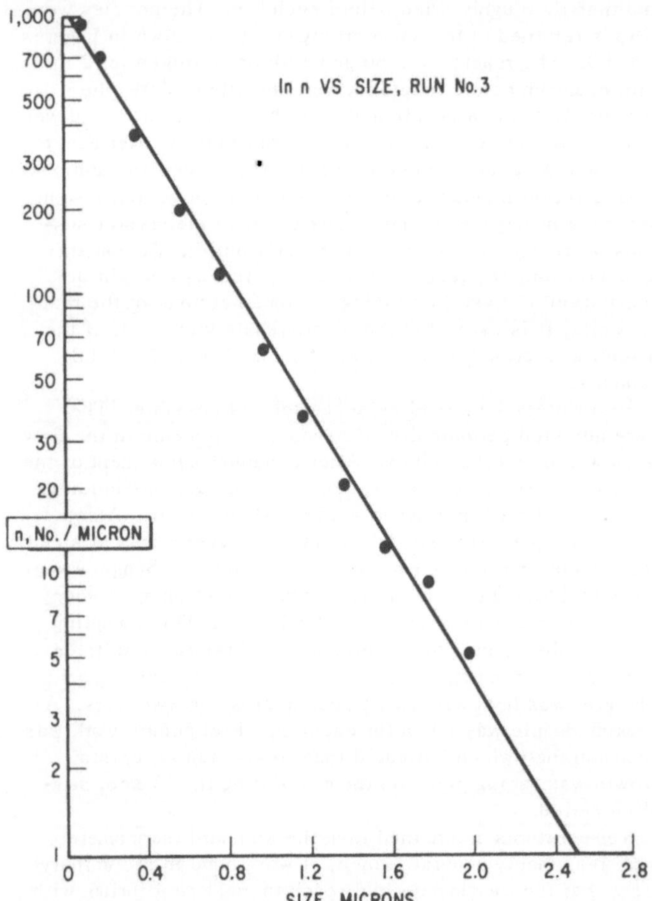

Fig. 4. Population density vs. size from continuous ammonium chloride crystallizer, run No. 3.

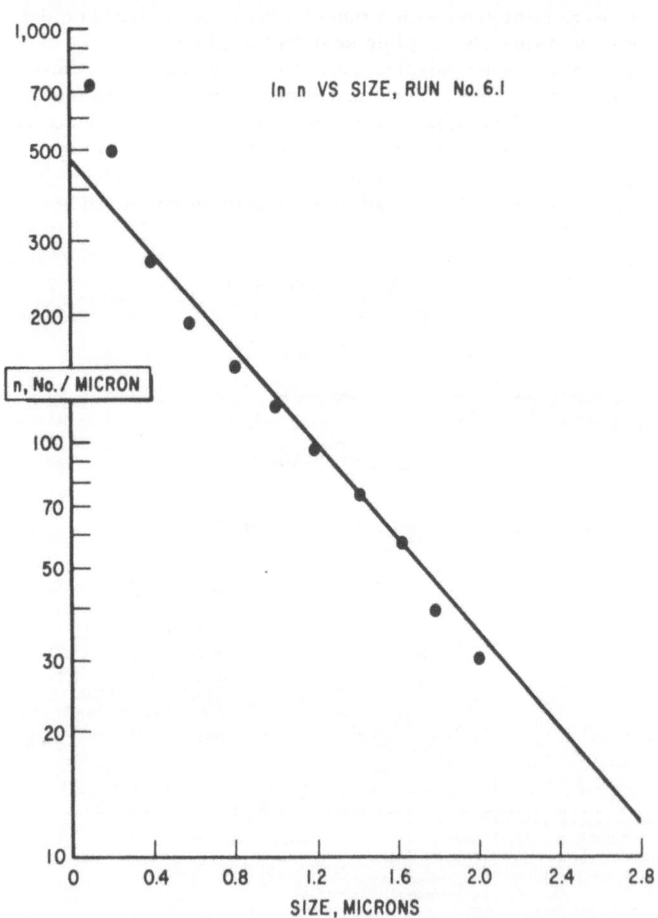

Fig. 5. Population density vs. size from continuous ammonium chloride crystallizer, run No. 6.1.

of X2,000 because it was a good compromise magnification for the total crystal size distribution produced and for the distribution field shown in the electron micrograph.

RESULTS

CRYSTAL SIZE DISTRIBUTION

One of the objectives of this study was to see if the theoretical CSD as derived from the population balance was observed for crystals produced in a gas-phase, backmixed environment. As predicted, the experimental CSD for the MSMPR reactor did follow a straight line as shown in Figure 1. The next two graphs (Figures 4 and 5) are average examples of experimental CSD data that was obtained.

Overall, in thirty-one experimental data runs, twenty-six of them exhibited the straight line CSD as shown. As a corollary to this observation, one may deduce that McCabe's ΔL law does apply in this system.

The crystal size distribution data referenced above was obtained by taking crystal samples, by counting and sizing the crystals into appropriate size ranges, by dividing the number in each range by the size range in microns, by smoothing these values, and then by plotting n vs. size on semilog coordinates. Smoothing the data consisted of plotting n vs. size on regular x-y coordinates and then drawing a smooth curve through the data points. An example of smoothed data is shown in Figure 6. This was justified, since the crystal samples were very small compared with the total number, and the distinct counting ranges could cause discontinuities in the exponential curve of the CSD if the sample of a particular range was not representative. However, in about one-half the data runs, the CSD data did not need to be smoothed.

The average crystal size range of ammonium chloride crystals produced throughout this study was from 0.1 to 2.5 μ. The only question arising in the crystal population density data was in the size range of 0.0 to 0.1 μ. In approximately one-third of the data runs, this size range usually contained too many crystals with respect to the straight-line CSD correlation. However, deviations in the data below 0.1 μ can be expected when one realizes the limitations of sizing, counting, and sampling techniques when working in this size range. Figure 5 illustrates a severe deviation in this smallest size range.

GROWTH RATES

Shown in Table 1 is the experimentally determined growth rate data. These growth rates were determined from the slope of the semilog straight-line plots as shown in Figure 1. Since the crystals were spherical solid types and spherical dendritic types, the growth rates shown are based on the overall spherical diameters of the crystals. As would be expected, the dendritic type of crystals usually have larger growth rates. Excess hydrochloric acid was used in runs No. 4 through No. 7 because preliminary tests indicated that this con-

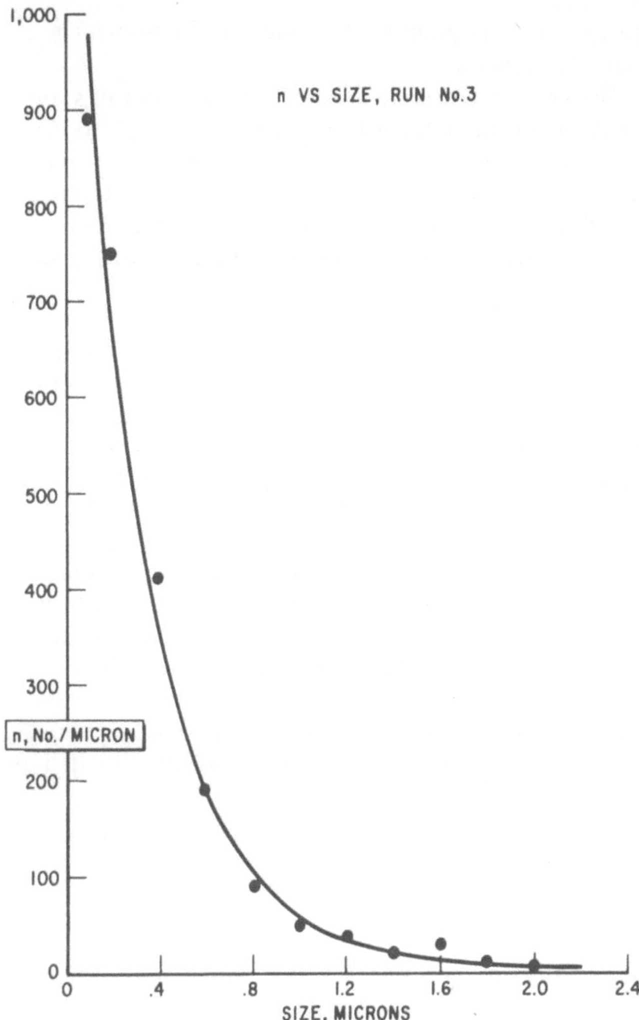

Fig. 6. Smoothing data.

TABLE 1. GROWTH RATES

Run No.	Holding Time sec.	Growth Rate microns/minute	Reactants cc./min. HCl	Reactants cc./min. NH_3	N_2 cc./min.	General Crystal Type	Straight-line Correlation
1.1	10	3.18	350	450	6000	solid	yes
1.2	10	4.82	350	450	6000	dendritic	yes
2.1	10	3.38	350	400	6000	dendritic	yes
2.2	10	2.08	350	400	6000	solid	yes
3	10	2.16	400	400	6000	solid	yes
4.1	8.6	2.25	225	200	7000	dendritic	yes
4.2	10	2.51	225	200	6000	both - 50/50	yes
4.3	12	1.97	225	200	5000	dendritic	yes
4.4	15	1.70	225	200	4000	dendritic	yes
5.1	8.6	4.11	225	200	7000	solid	yes
5.2	10	3.92	225	200	6000	dendritic	yes
5.3	12	3.36	225	200	5000	dendritic	yes
5.4	15	--	225	200	4000	dendritic	no
6.1	7.5	6.10	225	200	8000	dendritic	yes
6.2	8.6	6.07	225	200	7000	dendritic	yes
6.3	10	2.32	225	200	6000	solid	yes
6.4	12	2.25	225	200	5000	both - 50/50	yes
6.5	15	2.11	225	200	4000	solid	yes
6.6	20	1.11	225	200	3000	solid	yes
7.1	7.5	--	225	200	8000	dendritic	no
7.2	8.6	--	225	200	7000	dendritic	no
7.3	10	--	225	200	6000	dendritic	no
7.4	12	--	225	200	5000	dendritic (filled)	no
7.5	15	5.46	225	200	4000	dendritic	yes
7.6	20	2.25	225	200	3000	solid	yes

dition tended to produce the solid type of ammonium chloride crystals.

The next plot (Figure 7) shows the growth rates as functions of the holding time. Runs No. 5 and No. 6 performed overall as one would expect with decreasing growth rates for longer holding times. The fact that the growth rate data was not reproducible cannot be explained except for the discontinuities in crystal type.

NUCLEATION

By means of heat balances on run No. 6, the moles of ammonium chloride produced per cubic centimeter of outlet gases M_T were calculated as shown earlier in this paper. From these values of M_T an experimental $\bar{n}^\circ$ was calculated. Correlation of $\bar{n}^\circ$ vs. r indicated a semilog trend of higher nuclei density at the lower growth rates, a result which does not agree with the simple power model form. As may be seen from Figure 8, the data were not conclusive. Further analysis of this data was not warranted, since the same type of crystal was not produced in each subrun. However, in principle, the method does appear applicable to the determination of quantitative nucleation-growth rate kinetics.

Heat balance calculations (see Table 2) indicated an essentially constant yield of ammonium chloride in all the experimental subruns of run No. 6, with only approximately 10% of the reactants ever combining. No partial pressure measurements were made to check these thermochemical calculations. As calculated nuclei densities depend on this calculated degree of reaction [see Equation (11)], it is possible that erroneous values of $\bar{n}^\circ$ were calculated. It is difficult to conceive of such low yields of ammonium chloride in such an apparently fast reaction. The heat balance results are only suggestive and should not be interpreted as being quantitative values for this system. However, it is possible that diffusion of ammonia and hydrochloric acid through a stagnant film of nitrogen was the rate limiting step in

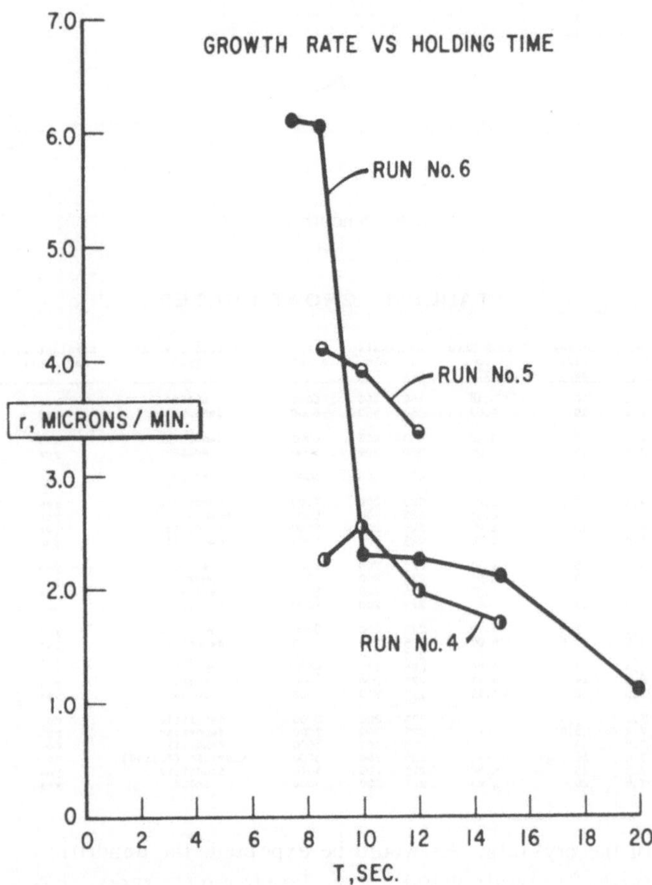

Fig. 7. Effect of changing holding times on the crystal growth rate of ammonium chloride in gas phase.

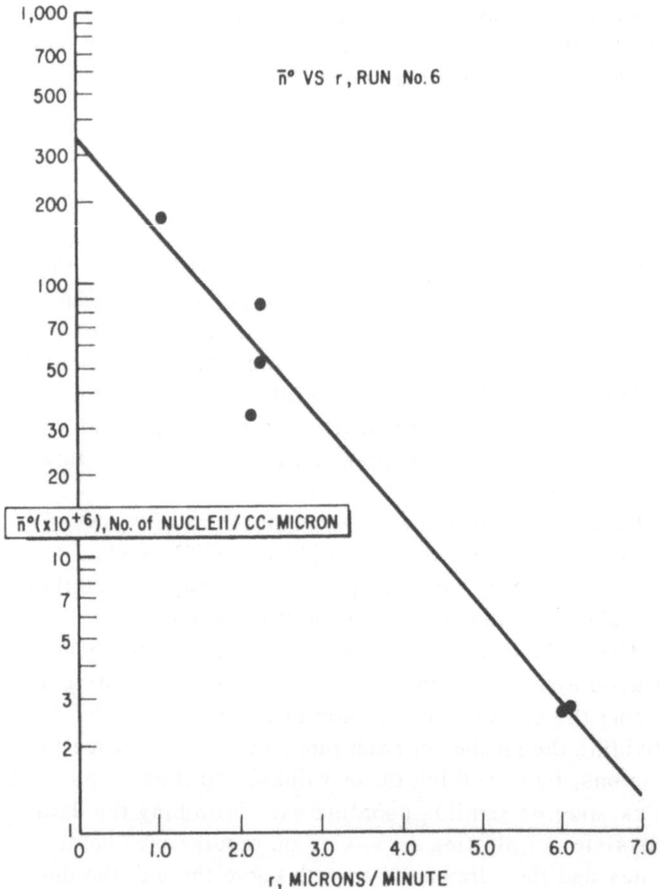

Fig. 8. Correlation between nuclei density and growth rate for crystallization of ammonium chloride.

TABLE 2. NUCLEATION DATA FROM RUN NO. 6

Sub-Run	Equil. Temp. °F	Approx. Q cal./min.	R_o moles/min. x 10^{-4}	$\bar{n}^\circ$, number/cc.-micron x 10^6	dN°/dt, number/cc. min. x 10^6
1	86	27.4	6.5	2.7	16
2	96	40.8	9.7	2.7	16
3	100	40.9	9.7	78.4	180
4	105	42.1	10.0	52.2	118
5	110	39.8	9.4	32.6	69
6	117	39.3	9.3	177.5	197

the reaction, thus causing the low yields. Further studies should include partial pressure measurements as well as CSD data. Time and equipment limitations prevented the gathering of such auxiliary data.

Much experimental work on nucleation phenomena as discussed above remains to be done. The limited data revealed little about nucleation kinetics. Although these results are far from the original objectives, the work is a beginning on the much needed data under conditions of a mixed crystal suspension.

CRYSTALLIZATION

One of the important problems encountered was that two types of crystals were produced by the reactor under the same operating conditions: solid crystals and dendritic crystals. Figures 9 and 10 show examples of these two types of crystals. Determining meaningful growth rates for comparison on the basis of overall crystal diameters was very difficult with two entirely different crystal forms. The erratic behavior in producing these two crystal types was investigated in eighteen special data runs, and no regular pattern or trend could be established in relation to the operating conditions.

Both the spherical solid crystals and the spherical dendritic crystals are quite characteristic of this ammonium chloride system. The spericity of these crystals was established by means of special electron microscopy shadowing and focusing techniques. The form and structure of ammonium chloride depend greatly upon the temperature, pressure, and impurities (*3*). Many variations in the crystal structure are possible. Any metallic chlorides are especially effective in modifying ammonium chloride. Buckley stated that ammonium chloride is addicted to a branching growth in three dimensions (*1*). These branches are always at right angles to one another. The pattern is of a rectangular (three-dimensional) branched configuration.

X-ray diffraction and electron diffraction analysis techniques were used to study many crystal samples from the reactor. Diffraction results indicated that each spherical particle in the sample micrographs was essentially a single crystal rather than an agglomeration of crystals. Therefore, each solid crystal was a single crystal, while each small limb of a dendritic crystal was the beginning of a larger perfect single crystal. If the vacant portions of a dendritic crystal were filled in with ammonium chloride molecules, the dendric crystal could become a whole single crystal. The directions of the crystal planes were not studied in this effort.

A possible overall mechanism for the reactor was concluded from this study. Dendritic crystals are formed in the gaseous phase. As they deposit on the reactor walls, they continue to grow to solid crystals. Since the carrier gas swirls and backmixes through the reactor at a high speed, solid crystals from the walls are carried along into the outlet stream at sporadic intervals. The

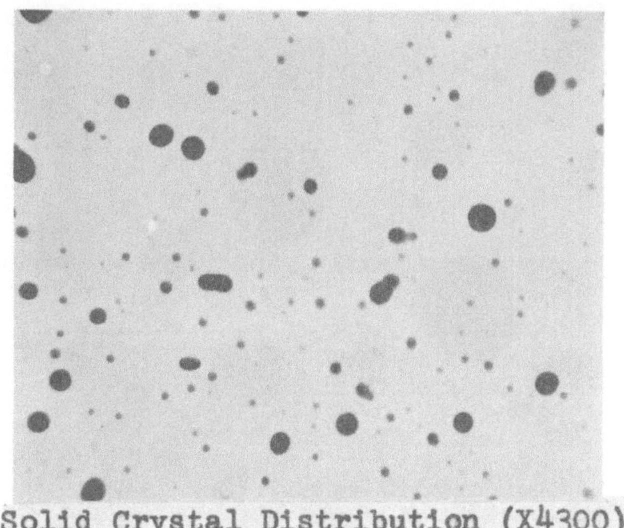

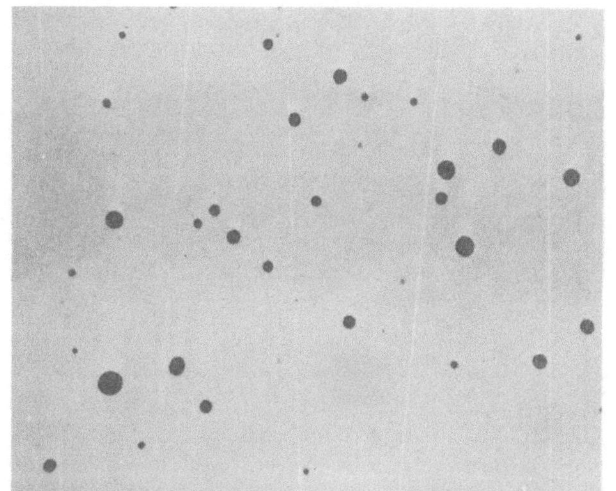

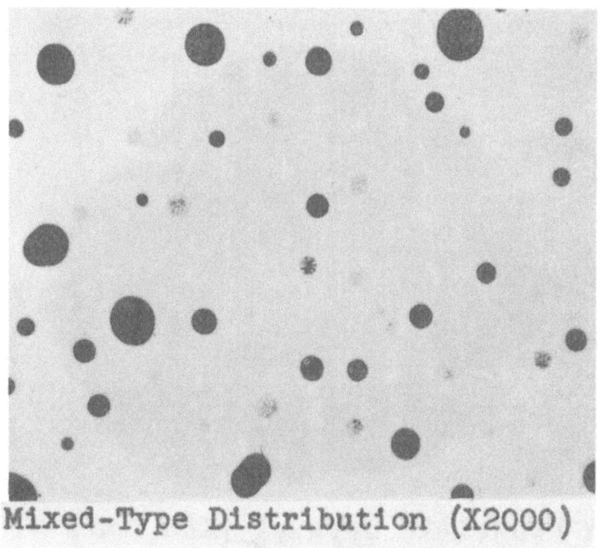

Fig. 9. Examples of crystal types produced in continuous reactor.

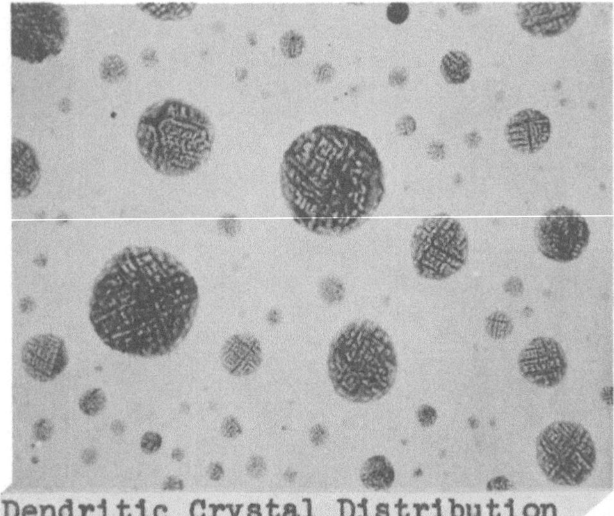

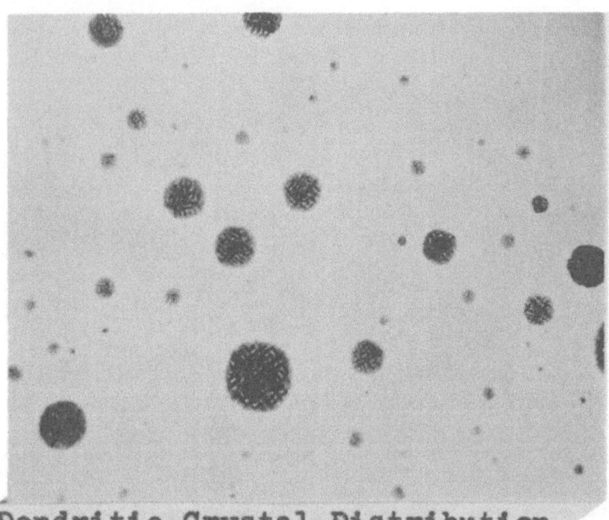

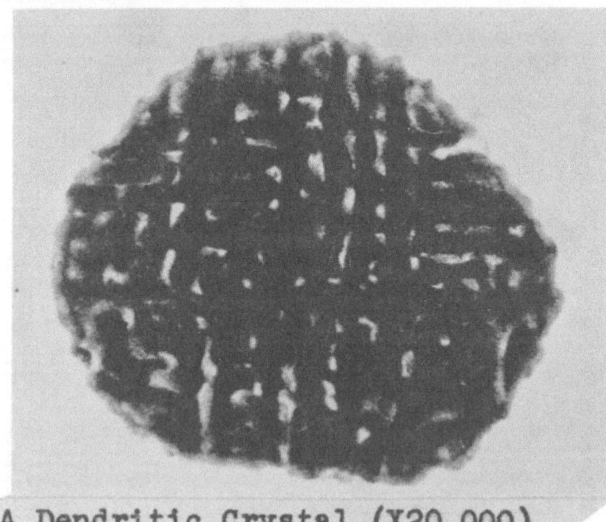

Fig. 10. Examples of crystal types produced in continuous reactor.

latter phenomenon is especially prevalent as the wall layer gets thicker. Therefore, solid crystals sometimes dominate the outlet flow as an unsteady state disturbance. This is a possible overall mechanism based on the limited data obtained. A much more thorough investigation is needed in order to adequately identify the crystallization phenomenon taking place. Quantitative nucleation-growth rate kinetics should only be measured when a particular crystal form was predominant.

CONCLUSIONS

Although the results presented in this paper are far from representing a complete picture of gas-phase crystallization, the authors believe that the experimental approach and techniques discussed are a valuable beginning toward studying the many critical phenomena involved in such a process. As discussed by Randolph (6), data from the MSMPR case can be effectively used to predict CSD in other modes of operation.

It is concluded that the theoretically predicted exponential form of CSD derived from a population balance does appear to agree with the authors' experimental gas-phase data. Also, the capability for obtaining crystal nucleation-growth rate kinetics by means of CSD data in a gas-phase, isothermal reactor-crystallizer producing very small particles was shown to exist. The population balance approach together with such techniques as electron microscopy and electron diffraction should prove to be a valuable means of studying gas-phase crystallization phenomena; however, additional chemical measurements, for example, partial pressures of the reactants, are necessary to formulate the kinetic model. Additional analytical work by the authors not discussed in this paper has indicated that such experimental data could be obtained with the ammonium chloride system; however, different gas-phase crystallization systems would require different methods of chemical sampling and analysis.

This experimental study serves as a source for many questions. Future work to answer these questions and clarify many areas should include such approaches as directly measuring the partial pressures of the reactants; obtaining similar data over a large range of nitrogen carrier gas dilutions; premixing the hydrochloric acid with nitrogen and the ammonia with nitrogen before they enter the reactor; evaluating different mixing geometries by means of nozzles, reactors, and baffles; investigating further the critical problem of producing two types of crystals under identical mass and energy conditions; and utilizing fully the capabilities of electron and X-ray analysis methods to study the crystallographic mechanisms of the crystals.

ACKNOWLEDGMENT

This work was accomplished at the Air Force Eastern Test Range Chemical Laboratory (ETOIL), Patrick Air Force Base,

Florida, while Captain Stone served in the United States Air Force.

NOTATION

$\overline{A}$ = area (sq. cm.) of crystals per unit volume (cc.) of suspension
k_a = crystal area shape factor, area/crystal-(length)^2
k_n = specific rate constant in kinetic nucleation rate equation
k_v = crystal volume shape factor, volume/crystal-(length)^3
L = crystal size or diameter, μ, measured along characteristic axis
M_T = total mass of crystals in suspension, moles/cc.
m = molecular weight
N_T = total number of particles
ΔN = total number of crystals in size range ΔL
N° = total number of nuclei
n = total crystal population density, number/μ
$\overline{n}$ = point crystal population density, number/cc.-μ
$\overline{n}_i$ = suspension input population density, number/cc.-μ
$\overline{n}_0$ = suspension output population density, number/cc.-μ
$\overline{n}_s$ = crystal population density at suspension surface, number/cc.-μ
n° = total nuclei population density, number/μ
$\overline{n}^\circ$ = nuclei population density, number/cc.-μ
p = partial pressure, mm. Hg
p_1, p_2 = partial pressures of ammonia and hydrochloric acid, respectively
Q = total heat from reaction
r = crystal growth rate along characteristic crystal axis, μ/min.
R = suspension input, or output, rate, g.-moles/min.
T = drawdown time, or holding time, V/Q, sec.
t = time, sec.
V = total suspension volume, cc.
$\overline{V}$ = volume (cc) of crystals per volume (cc.) of suspension
ρ_c = crystal density, g./cc.

LITERATURE CITED

1. Buckley, H. E., "Crystal Growth," Chapman and Hall, London (1951).
2. Bunn, Charles, "Crystals: Their Role in Nature and Science," Academic Press, New York (1964).
3. Kozlovskii, M. I., *Pedagog. Inst. im. T. G. Shevchenko*, **4**, 109 (1957).
4. Randolph, A. D., and M. A. Larson, *AIChE J.*, **3**, No. 5, 639 (1962).
5. Randolph, A. D., *Can. J. Chem. Eng.*, **42**, 280 (1964).
6. ———, *AIChE J.*, **11**, No. 3, 424 (1965).
7. Stone, P. D., M.S. thesis, Univ. Fla., Gainesville (1967).

CONTINUOUS SUGAR CRYSTALLIZATION: A CHEMICAL ENGINEER'S VIEWPOINT

Richard C. Bennett

Although sugar crystallization is one of the oldest and most important crystallization activities, it is still accomplished by batch methods. A chemical engineer has much to offer in possible improvements in the current process and in the design of possible continuous processes. A workable continuous process is described, along with a discussion of the economics engendered.

From the standpoint of a chemical engineer, the crystallization of both salt and sugar are by far the oldest and probably still among the most important of crystallization activities. The crystallization of sugar was known in prehistoric times, and records from 500 A.D. in Persia show that people collected syrups and boiled them to produce a purified form of crystalline sugar. This same general procedure seems to have been in use for a thousand years or more until roughly the beginning of the nineteenth century, when a French inventor developed the vacuum pan. The use of vacuum evaporation permitted operation at temperatures much lower than had been the practice previously and led to less discoloration of the syrups and solutions of much lower concentration from which the separation of crystalline material was easier. The simultaneous development of the centrifugal separator, or centrifuge, about 1837 permitted separation of the adhering mother liquors to a degree not previously achieved, and within the next 50 yr. or so the development of sugar crystallization as it is practiced today came into being.

It is worth noting that the crystallization of sugar is unique from the standpoint of the production of organic chemicals. The amount of sugar crystallized per year is equivalent to an average consumption of approximately 100 lb./person/yr., or about 27,000 tons/day.

Even more amazing from the standpoint of engineers acquainted with modern processing technology is the fact that all of this sugar is produced by batch crystallization techniques, the fundamentals of which were well known, understood, and in practice at the turn of the twentieth century.

The flowsheet of modern cane and beet sugar refineries has been so well publicized over the years that there is little point in reviewing other than a few of the basic steps which are involved prior to the crystallization step. Cane sugar is cut and transported at the appropriate time to a mill, where a primary extraction is made to separate the sugar from the stock and other natural foliage. This raw juice is concentrated in evaporators and then crystallized in raw sugar pans. The solutions from which the sugar is crystallized are quite dark and relatively viscous, and the crystalline material called *raw sugar* is a relatively large crystal of a brown color which caused by adhering molasses. The raw sugar is taken into a refinery where it is washed and dissolved (melted). The dissolved sugar is treated and filtered to reduce impurities and then concentrated to about 70 to 75% sugar with a purity of roughly 99.5%. The solution is boiled until a supersaturated condition, as measured by boiling point elevation, exists; then seed crystals are added. Depending upon the quantity of the seed crystals and their size, the pan may be either shock seeded or fully seeded. Boiling of the pan is done in such a way that nucleation is minimized, and the production rate is increased as the seed surface available for deposition of new growth increases. Although many techniques are used, in general operation is in such a way that the supersaturation coefficient (S) within the system is held somewhere between 1.1 and 1.4. The end of the batch is sensed by a lack of mobility of the magma. During the boiling process a high delta T exists between the steam chest and the slurry in

Swenson, Division of Whiting Corporation, Harvey, Illinois.

the pan. Although most pans employ mechanical circulators, there are many which rely solely on natural circulation. The viscosity of a supersaturated solution ranges from about 100 to 300 centipoise, and the magma density at the end is 50% crystals by weight or about 70 to 90% by true volume.

At the end of the batch a large valve in the bottom of the pan is opened, and the entire mass is dumped into an agitated, jacketed receiving vessel called a *mixer*. Refineries vary greatly on the exact length of time used for a strike, but a typical fast strike would equal about 70 min. Some California refineries make longer strikes to achieve better grain formation. A typical long strike is 110 min. Listed below are the typical times allotted for various functions:

	Typical	Long
Fill	5 min.	7 min.
Boil	7 min.	11 min.
Ripen	8 min.	35 min.
Feed liquor	39 min.	40 min.
Stiffen	4 min.	4 min.
Dump	3 min.	3 min.
Steam out	4 min.	5 min.

Beneath the mixer are two or more centrifuges generally operating on a programmed batch cycle, and below these are mother liquor storage tanks. The system is vertically disposed, and the evaporators are generally located on the fourth or fifth floor.

The purified feed at 99.5% has about one-half of the total sugar available removed in the first strike. This decreases the purity level to 99% for the mother liquor feed to the second strike pan, where again 50% of the total sugar is removed and the purity level decreases to 98%. After the third strike the purity level drops to 96%, and after the fourth strike it drops to 88%. The mother liquor leaving the centrifuge after the fourth strike is called *fourth runoff liquor.* Some refineries make a fifth strike, and liquor from this strike is called *fifth runoff liquor.*

Sugar from the second strike is generally the same as that shipped as product. Sugar from the first strike is blended with enough third and fourth strike material so that its color equals that from the second strike. In other words, the purity of the average product is less than that of the first strike and more than that of the third or fourth strikes. Generally speaking, some of the fourth strike and all of the fifth strike sugar is too dark for use and must be reprocessed into remelt sugars.

Since the bulk of the sugar is produced in the first strike, the quantity of material available for feed to the succeeding strikes is much smaller than the incoming feed. Based on a 50% yield in each strike, the following are the calculated yields per 100 lb. of white pan feed:

Solution	Feed	First Strike	Second	Third	Fourth	Total
Sugar (lb.)	75.0	38.0	19.0	9.5	4.75	—
Impurities (lb.)	0.38	0.38	0.38	0.38	0.38	—
Water (lb.)	24.62	9.22	4.82	2.47	1.27	—
Purity	99.50	99.0	98.0	96.1	92.6	—
Product crystals (lb.)	—	37.0	19.0	9.5	4.75	70.25
Evaporation (lb.)	—	15.4	4.4	2.35	1.20	23.35
% of yield	—	52.8	27.0	13.5	6.7	—
% of evaporation	—	66.1	18.9	10.0	5.0	—

The amount of evaporation and production during the first strike is larger than the sum of all the rest. This requires storage of mother liquor from first and higher strikes until there is sufficient material available and equipment available to perform the higher strike. It also requires surge capacity for the products from these strikes so blending can be done for final product color.

Not unexpectedly the physical plant which houses a sugar crystallization system and supplies its utilities has been tailored over the years to meet the specific requirements of this process. From the very earliest days it has been the practice to handle the magmas from a sugar crystallizer by gravity. Large foot valves are used to dump the batch crystallizers into mixers or agitated vessels which hold the slurry prior to batch centrifugation in suspended basket centrifugals. The liquors leaving the centrifuges run by gravity to tanks at a lower elevation, and, therefore, the entire process is generally vertically disposed with the crystallizing equipment being located on the third or fourth floor of the building. This leads to much larger and sturdier structures than are typical for most processing plants involving similar equipment.

Of even greater interest, however, to chemical engineers are the implications of a batch process on the heat and material balance around a sugar crystallization system. The filtrate from the centrifugals is at saturation or slightly supersaturated and contains some fine solids as well as wash water added to purify the crystals during centrifugation. The mixture of this material plus dilution water required to drop the level of supersaturation back to the unsaturated condition prior to commencing the next strike and also to permit filtration or other purifying steps between the first and second strike results in a solution that may be as low in concentration as 60 to 65% according to some Eastern producers. On this basis the total steam usage for the refined sugar production would be about 82.4 lb./100 lb. of crystals for a batch system, whereas if the final mother liquor purity were the same for a continuous system, only 52 lb. of evaporation would be required. In addition, this steam comparison does not include the steam that is required in a batch system to steam out the pans at the conclusion of each batch, pull vacuum on the new batch, or the steam required to heat the syrups between batches.

This difference between the steam requirement for a conventional batch process and the potentially lower steam consumption of a continuous process is a fundamental difference and is a significant quantity. On the basis of steam worth 75¢ a thousand lb. and 8,000 hr./yr. operation, this difference in steam consumption would be approximately $150,000 /yr. for a 1,000 ton/day plant. In addition to this, there is a significant amount of difference between the peak steam load required with the conventional batch pans and the continuous or steady steam demand of a continuous process. According to information in Honig, the peak steam demand shown for a typical batch operation is 42,000 lb./hr., whereas the average rate over a 2-hr. cycle is only 22,500 lb./hr. This places the peak demand at 187% of the average value. When we assume that these figures are representative of a refinery's operation, it is obvious that by going to a continuous process a larger production rate could be realized from the same equipment, assuming the steam power plant is sized for the peak demands. This is potentially a very important way of increasing capacity without substantially increasing capital cost.

It certainly would seem strange to modern chemical process engineers that a basic industry like the sugar industry which has been in profitable operation for longer than probably any other segment of the chemical industry should continue to use processing techniques that by most standards require higher utility consumption, a greater amount of highly skilled labor, and greater capital investment than the continuous processing technology which has become so popular in the last 20 or 30 yr. As would be expected, a number of people have tried rather unsuccessfully to develop or invent continuous sugar processes that would give the same quality of production that the current batch process produces.

The major reason for producing a crystalline sugar is to separate a high purity sucrose for a solution containing other sugarlike materials generally described as molasses. In addition to its purity, the size of the crystal, its ash content, and its physical appearance are also important properties to any producer. In this separation process the forming of a crystal without occlusions of mother liquor is important. Just as important is the shape of the crystal, since the solutions tend to be viscous. Typical viscosities range from 50 to 300 centipoise. If the crystalline sugar tends to be agglomerated or exhibits dendritic crystalline formation, then the washing of adhering mother liquor from the crystals in the centrifuge becomes a very difficult problem. Production of good quality crystals is possible in a batch process only when skillful operation is employed. Improper seeding, seeding at the wrong time, too high a level of supersaturation during the batch, or too short a growth period can lead to a faulty grain formation. The skill of a sugar boiler is high and demands premium wages and treatment.

It should be noted here that the crystallization of sugar is unique from the standpoint of the experience of people used to handling typical inorganic salts which support very low levels of supersaturation and nucleate profusely both from high levels of supersaturation or relatively high levels of mechanical stimulation. By contrast, sugar (and here I am referring to both beet and cane sugar) will behave in more a classic sense from the standpoint of the Oswald concept of solubility. When boiled, sugar will lose water and show very large boiling point elevations which are proportional to its supersaturation. These have been depicted in the literature (6), and a typical example is shown in Figure 1.

In addition to developing very high and easily measured degrees of supersaturation, the sugar solutions quite typically will exhibit a stable supersaturation. In other words, once developed, this level of supersaturation does not degenerate quickly into nucleation but appears to remain quite stable for a matter of minutes or even longer. This supersaturation is also stable in the presence of considerable mechanical stimulation. A beaker or agitated vessel filled with supersaturated solution and crystals may be stirred by relatively high-speed stirring devices without a substantial amount of new nuclei being created from these mechanical effects. From the comparative standpoint, sugar will tolerate degrees of mechanical stimulation that would induce profuse nucleation in such materials as sodium chloride solutions, potassium chloride solutions, or other typical inorganic salts.

Because of this characteristic lack of nucleation, sugar crystallizers have for many years been operated in a batch manner with the introduction of seed crystals which constitute either all of the seeds required for the

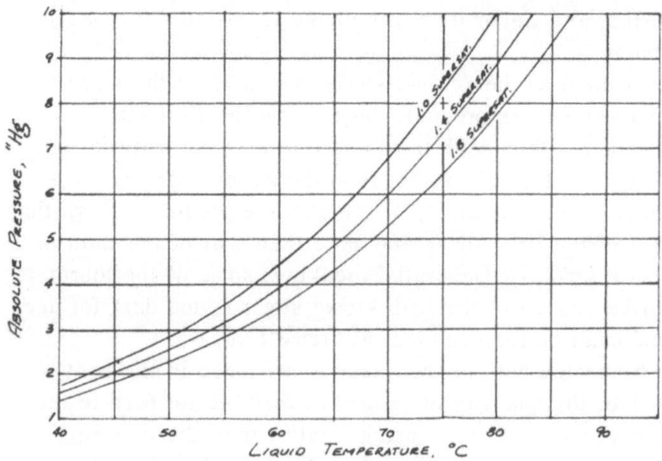

Fig. 1. Boiling elevation.

final magma or part of them, depending on the level of initial supersaturation at the time of seeding. This seeding tends to reduce the coefficient of variation in the product produced from a strike pan from the characteristic coefficient of variation expected of roughly 55% for a fully mixed continuous crystallizer down to a range of around 30 to 40%. This narrower coefficient of variation represents less oversize and less fine material than would be produced from a continuous process. Although many different sizes of sugar are produced for human consumption (and this is virtually the only market for sugar), typical practice is to vary the average particle by the length of the boiling time and the amount of seeding but still produce a relatively narrow size distribution through the normal strike pan techniques. This narrower size distribution also is helpful in the centrifugation of the product crystals.

DEVELOPMENT OF A CONTINUOUS SUGAR CRYSTALLIZER

In the early 1960's, the Swenson Evaporator Company, which got its start in the 1890's through the supply of numerous beet sugar evaporative crystallizers in the South and Western parts of the United States, undertook some development work in the field of continuous crystallization of sugar. This work ultimately led to the development of a continuous crystallizer of the forced circulation type capable of producing crystalline sucrose from both beet and cane sugar under a variety of operating temperatures, conditions, and solution compositions.

Early investigative work led to the conclusion that a continuous sugar crystallizer operated in the conventional unseeded mode, which is characteristic of most inorganic salt crystallizations, would lead to a self-seeded system capable of producing a product that ranged from about 20 to 50 U.S. mesh, depending on the purity of the mother liquor in which crystallization took place. The product from this continuous crystallizer which could be maintained at supersaturation coefficients ranging from 1.10 to 1.8 was of strikingly good conformation and appearance, low in ash content, and low in color content. Attached are photographs of screened portions of the product produced in one of these runs (Figure 2) and also photographs of typical commercial sugar production (Figure 3). The physical properties of a sample of continuously produced sugar and the typical batch material are shown in Table 1.

The size of the crystals produced from a system which is unseeded through any external mechanism and sustains itself through self-nucleation imparted to the system by the recirculation pump and piping turbulence tends to be somewhat coarser than most of the commercial material used for human consumption. In general, the product sold to the U.S. markets consists of material that ranges from 40 to 60 U.S. mesh in size. Somewhat finer sizes and ground sugar (powdered) are used for special

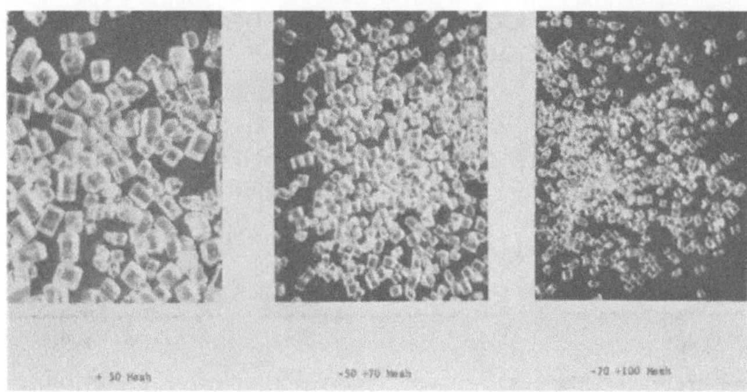

Fig. 2. Crystal product from continuous crystallizer.

applications, and some very granular material is used for other applications. The bulk of the product, however, is crystallized in the 40 to 60 mesh size range for use on U.S. tables.

The product size distribution of the crystalline sugar produced in a continuous crystallizer of the forced circulation type similar to that shown in Figure 4 running in an unseeded condition corresponds to a coefficient of variation of approximately 45%. This coefficient of variation is close to the theoretical expected for a fully mixed suspension, mixed product removal crystallizer of the type described by Randolph and Larson (2) and

Fig. 3. Crystal product from a commercial crystallizer.

TABLE 1. COMPOSITION OF BATCH AND CONTINUOUS SUGAR

	Typical Batch Product	SWENSON Continuous Product
Crystal Size-Tyler		
+20	0.1	5.5
+28	1.9	33.2
+35	21.2	30.2
+48	35.6	15.2
+80	34.1	10.3
-80	7.1	5.6
Mean Aperture (Microns)	326	524
C.V.	39.8	41.9
Vd. Weight - lbs/Ft^3	49.5	56.7
Color 1000a$_{420}$	8.9	8.1
Ash, % Conductivity	.0062	.0016
Invert Sugars	.009	.0057

Randolph (7) when operated at constant feed rate, constant suspension volume, and constant slurry density with a material whose growth follows the McCabe delta L law. Shown in Figure 5 is a plot of the population density vs. the crystal diameter for sample No. 55, indicating that the expected straight line is obtained and the calculated growth rate r is 0.041 mm./hr. The nuclei population density, based on 1 liter of slurry, is the number whose $\log_e$ is 19.6. In a similar manner these calculations have been made for a number of other samples which were taken when the system was at or near steady state conditions. These samples are shown in Table 2.

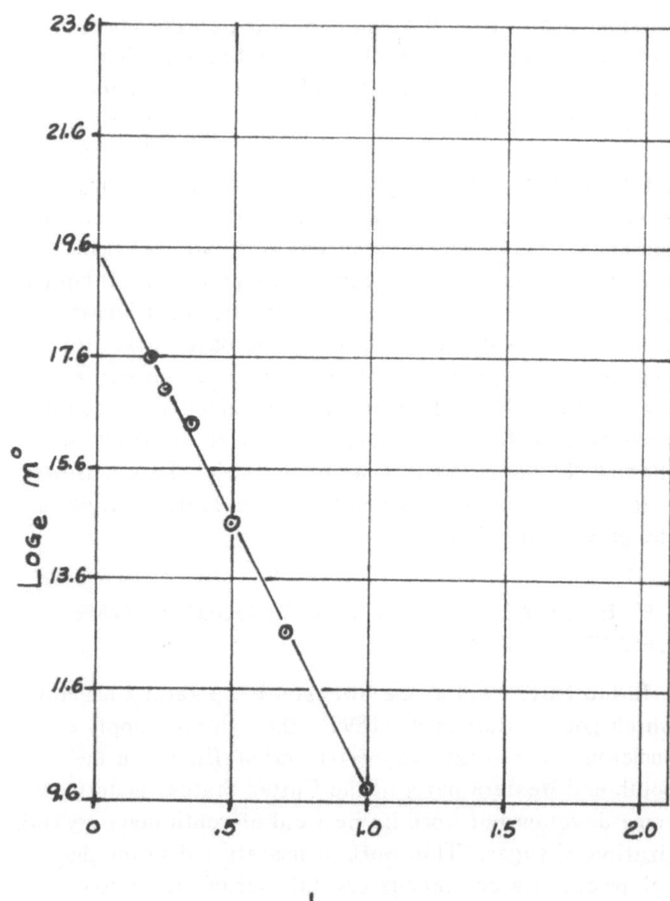

Fig. 5. Crystal size distribution.

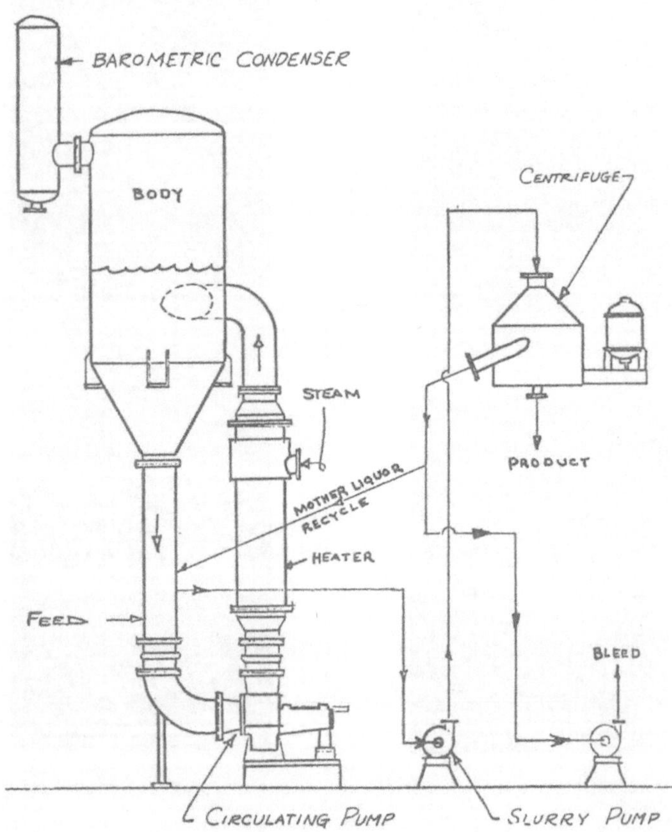

Fig. 4. Continuous crystallizer.

The samples were taken with ordinary commercial techniques, and the equipment was controlled with conventional air-operated commercial instruments. Most of the work was done at approximately 70°C. liquor temperature. In general, a continuous sugar crystallizer reaches stable equilibrium, although in this work the feed concentration was hand controlled, and it was difficult to maintain the slurry density at a constant value. The mathematical analysis of the samples in accordance with the techniques described by Larson was done at a much later date than the bulk of the work covered by this report, and in certain cases where it may have been desirable to expand upon the data to give a more complete theoretical coverage unfortunately, there was no opportunity to do so.

Shown in Figure 6 is the relation between growth rate and supersaturation coefficient for the samples in Table 2 with a solution of 100 purity. The supersaturation coefficient is that used in Gillette's work and is obtained from the boiling point elevation as shown in Figure 1. The growth rate of 0 mm./hr. at a supersaturation of 1.14 may be due to short circuiting, an error in the relationship between the boiling point elevation measurements and

TABLE 2. EXPERIMENTAL DATA FROM PILOT PLANT CRYSTALLIZER

Calc. No.	r mm/hr	loge n°	Density gms/L.	Super Sat. Coe.	Solution Purity	Pump Speed RPM	Tip Speed Ft/Min	Retention Hrs.	Nuc. Speed Ft/M	% C.V.	M.P.D. Inches	+14	+20	+28	+35	+48	+65	Date
167	.114	17.3	200	1.5	100	920	2760	1.39	Off	43.5	.023	22	14	50	72	88	94.5	4/29/61
143	.085	17.8	186	1.45	100	920	2760	1.66	Off	45.3	.0215		13	45	68	85	93	
155	.159	16.4	390	1.80	100	750	2250	1.54	Off	44.0	.034	25	56	79	91	96	99+	11/2/66
156	.170	16.1	367	1.74	100	750	2250	1.51	Off	44.0	.034	20	53	79	91	96	99+	
54	.069	17.4	250	1.38	100	760	2280	2.5	Off	42.0	.025		24	61	77+	91.5	97	4/5/62
55	.041	19.6	335	1.29	100	760	2280	2.5	4600	48.4	.015		3	14	38	76	92	
50	.1205	15.2	75	1.30	100	650	1950	1.66	Off	59.5	.0310	23	49	68	82	99+		3/14/62
174	.1020	16.4	98	1.18	100	650	1950	1.66	2500	60.5	.024	11	30	53	70	93		
57	.0975	13.9	50	1.35	90	760	2280	2.70	Off	37.2	.035	20	59	82.5	94	99+		3/29/62
59	.087	15.9	160	1.28	90	760	2280	2.50	4600			6.9	37.6	70.4	89.2	98.2	100	3/30/62-G

the supersaturation for the solutions used, or it may be an indication of the critical supersaturation required to initiate growth, or a combination of both. No attempt was made to accurately define this area in the present work because the area of interest is at higher supersaturation coefficients where growth proceeds at relatively high rates. From a supersaturation coefficient of 1.2 to 1.8, linear relation was found between the supersaturation coefficient and growth rate.

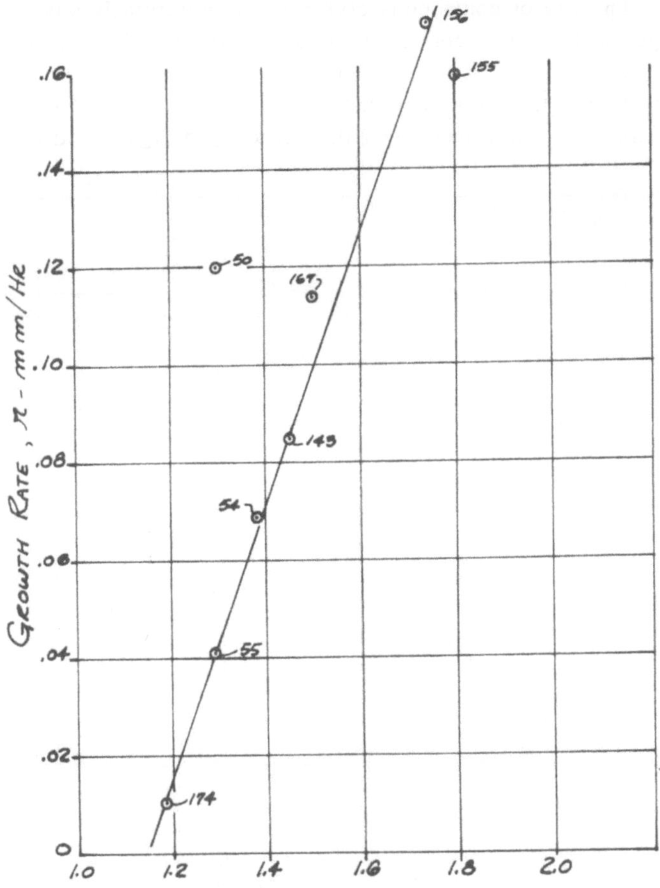

Fig. 6. Crystal growth rate.

Shown in Table 3 are the measured growth rates compared with values taken from the literature. In general, the values obtained are much lower, which may account for the better grain formation during these tests. One weakness of the data reported in the literature is that the method and conditions under which the measurements were made are not directly comparable with commercial practice, and most likely they are maximum values rather than realistic values for plant operation.

In general, the product made with only the nucleation caused by the circulation pump was insufficient to give the size product wanted for table use when purities are high. It was noted that the use of increased pump speeds leads to smaller product sizes, and the mechanism appeared to be increased nucleation due to mechanical stimulus of the supersaturated solution. The graph of

TABLE 3. GROWTH RATE DATA, MILLIMETERS PER HOUR

	This Work*	Honig[3]	Van Hook[4]	Chapman[5]
Temperature	70°C	70°C	30°C	70°C
S. Sat. Coe. (S)				
1.1			.030	
1.11				.096
1.175				.144
1.20	.0075		.080	
1.40	.0350		.200	
1.60	.0645			
1.80	.0910		.300	
S. Sat. not spec. in equivalent units.		.0322 to .151		

*Data of Figure 6 are in diametrial rates. The values above are rates per face–taken as the diametrial rate divided by 2.

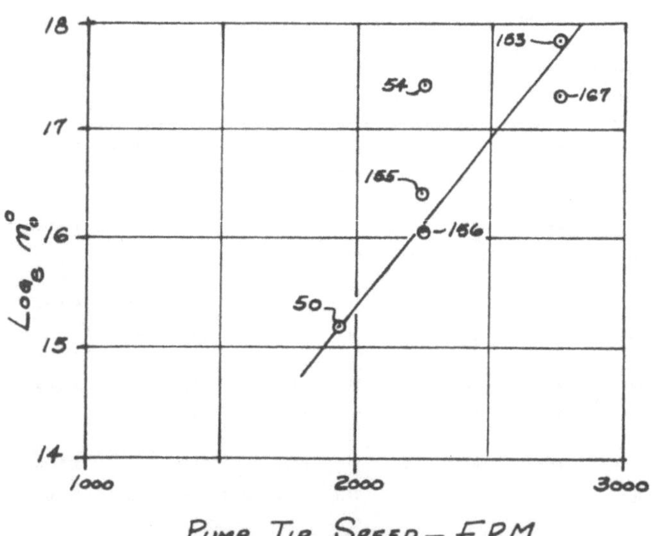

Fig. 7. Nuclei population density vs. tip speed.

$\log_e n_0^\circ$ vs. the tip speed of the circulation pump shown in Figure 7 indicates that this relationship may be a straight-line relation. Point No. 54 is an exception. If the average particle diameter is plotted vs. the tip speed, a linear relationship is also obtained indicating decreasing size with increasing speed. This is shown in Figure 8.

The use of increased mechanical stimulus to increase the rate of nucleation suggests the use of additional devices to add to the nucleation rate established by the pump, since in any real situation the use of increased pumping may be a relatively costly method of attaining this end. A number of devices were tried, but the simplest technique, and a very successful one, was to insert into the body of the crystallizer a circular saw blade which gave a high degree of turbulence when spinning at high speed in the supersaturated solution. The use of this device gave increased nucleation and reduced particle size as expected. The plot of $\log_e n_0^\circ$ vs. size retained the expected straight-line relationship (see Figure 5), indicating that the mechanism is nucleation rather than attrition due to impact with crystals in the slurry which would be expected to give a departure from a straight line due to the likelihood of shattering crystals into particles larger than a nuclei; hence, there is a deficiency in particles in the smaller sizes.

Shown in Figure 9 are the results at various pump and nucleator speeds. The straight-line relationship indicated is based on the expectation that intermediate speeds would give proportional values based on the straight line obtained in Figure 7, although this was not established experimentally. In the test work, mechanical limitations prevented operating the nucleators at a wide range of speeds.

From the data points No. 54 and No. 55 it can be seen that the influence of adding additional nucleation in the body reduced the average particle from 0.025 to 0.015 in.

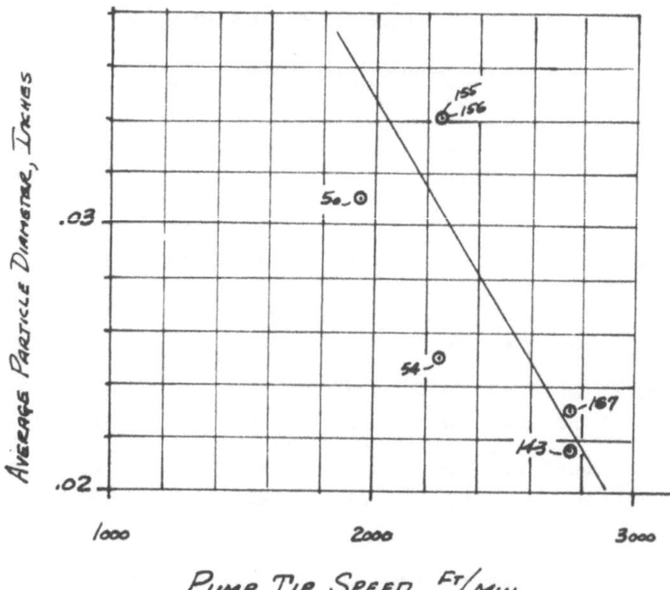

Fig. 8. Average particle diameter vs. tip speed.

The coefficient of variation remained essentially the same, at about 40 to 50%.

CONTINUOUS CENTRIFUGATION

The use of continuous crystallization naturally suggests the use of continuous centrifugation. The use of this technique has also been tried in the past and found to be wanting generally because of damage to the crystals by the movement over the screen and high speed un-

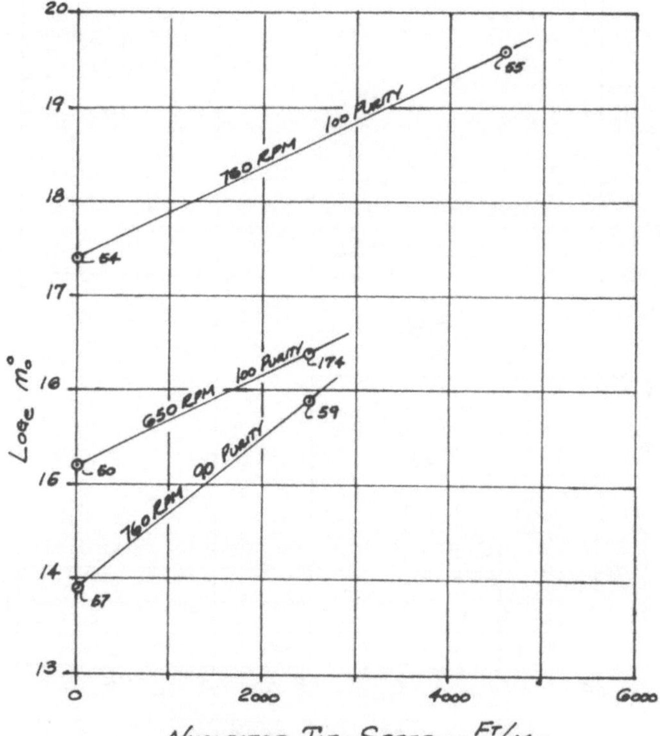

Fig. 9. Nuclei population density vs. nucleator tip speed.

loading, poor washing of the adhering mother liquor, and inability to handle the heavy slurry densities. Conventional practice is to use an automated vertical suspended basket machine which loads at low speed, spins at high speed with appropriate washing, and unloads at low speed to minimize damage to the product.

The use of the continuous crystallizer which operates at lower magma densities than the conventional batch pan naturally suggested the use of more modern centrifuge designs which can operate successfully at low slurry densities. The product made in the continuous machine is generally simpler and almost without the complex or agglomerated crystals seen from most batch pan operations. Such a product should be easier to wash and less susceptible to mechanical damage than the products which have been evaluated in continuous machines before.

During the continuous crystallization tests two styles of continuous centrifuges were tried, and both worked satisfactorily, with certain limitations. The pusher centrifuge was used with coarse material, similar to that described in calculations 155 and 156. When fed with a 25% weight slurry and operated with a 400-μ screen at speeds corresponding to G forces of 225 to 500, the residual moisture from the centrifuge ranged from 1.2 to 0.7%. These figures were at a wash rate of 0.07 lb./lb. of product. Indications are that the pusher design will give usable production rates with products down to about 40 mesh in size.

Of perhaps more interest is the worm-sieve machine similar to the Swenson-KMI model KMZ. This centrifuge has a cone shaped perforated screen whose openings can be obtained in various sizes so the action of the centrifuge can be made to separate product of a predetermined size range, while the remainder of the smaller crystals in the mother liquor are returned to the crystallizer for further growth. The action is then that of a wet classifier as well as that of centrifugation. This classification not only acts to seed the machine but also to reduce the coefficient of variation of the product discharged from the 40 to 45% produced naturally by self-seeding and/or mechanical stimulus to lower values, approaching 30%, depending on the rate of recirculation to the centrifuge and the size of the screen. While the worm-sieve machine is capable of handling low slurry densities, it operates with a relatively thin cake compared with the basket and pusher machines and, therefore, cannot give the product handled as thorough a washing as the former. To some degree, therefore, the simpler grain formation of the continuous crystallizer is required to utilize the continuous worm-sieve centrifuge.

A CONTINUOUS SUGAR CRYSTALLIZATION PROCESS

In view of the discussion above, it would appear that a continuous sugar crystallization process should make use of a proper centrifuge whose action can complement the continuous crystallizer, and such an arrangement is shown in Figure 4. In this arrangement the continuous crystallizer operating with or without an additional nucleating device discharges its product slurry to a classifying centrifuge with product crystals being removed and the fines in the mother liquor being returned to the body for additional growth. The return of seed in this manner would be a catastrophy in the crystallization of most inorganic compounds which tend to over nucleate, but with sugar the additional nuclei which may form from cooling of the returning mother liquor are helpful, since the system must be operated in such a way that nucleation is promoted. In any large-scale design a number of stages would be employed, and each would have its corresponding centrifuge (s) so that the product which was separated at each stage, which corresponds to a different purity level, could be handled separately for blending to get the final purity, color, and appearance desired. As shown in the attached data in Table 2, sufficient work was done at lower purity levels to establish the required retention times and the effectiveness of the nucleator as a method of size reduction.

A LARGE SCALE PLANT DESIGN

The concepts embodied in the foregoing paragraphs have been translated into a large-scale design in an effort to establish the overall costs, utilities, and operating labor requirements for comparison with existing plants. The selling price of the crystallizers for a three-stage design capable of 1,000 tons/day production would be approximately $160,000 in steel construction. The connected horsepower for the circulating pumps would be approximately 400 hp. The selling price of the same equipment in type 316 stainless steel with cast iron circulating pumps and 316 stainless trim would be about $260,000.

The equivalent production rate for a batch system is about 30 lb./hr./cu. ft. and the horsepower for a 1,600 cu. ft. pan is about 75. For an equivalent production rate, two such pans would be required, and the estimated price in stainless clad construction would be about $240,000. On this basis the cost of the three-stage continuous machine is slightly greater than the batch unit, although the number of continuous machines could be reduced if it were possible to field fabricate the vessels, since the two stage unit would require bodies larger than 12 ft. 0 in. The connected horsepower of the continuous unit is also higher than the batch system which relies to a large degree on natural recirculation.

On the installed basis the continuous equipment would be significantly less, since the vessels do not have to be supported at as high an elevation as with the batch system, the amount of tankage is less, and the liquid transfer and piping can be smaller than on the batch equipment. The most important differences, however, in using continuous equipment rather than batch equipment do not involve differences in capital cost but may be classified as utility savings, labor savings, and operational advantages.

UTILITY SAVINGS

The expected steam usage for a continuous crystallization system is 52 lb./100 lb. of product vs. 80 to 90 in the present batch equipment (based on dilution of the mother liquor to 65% between strikes). Based on a calculated value of 82.4 for the batch system, this is equivalent to about $150,000/yr. savings in steam for a production rate of 1,000 tons/day. This calculation is based on steam at 75¢/1,000 lb.

The peak steam consumption is the same as the steady state design value in continuous equipment. In batch crystallization the peak steam and water demand are about 187% of the average rate. In new plants this would permit the use of smaller and less expensive utility supply systems, electrical systems, water pumps, etc., or in modifying existing plants this means there is opportunity for increased production rates without modification of the existing boilers and ancillary equipment. This also means that a more suitable steam power balance can be achieved with a continuous system when compared with the existing batch systems.

Associated with the batch operation are relatively large air ejectors and complex direct contact condensing systems designed to meet the fluctuating load which is characteristic of batch operations. This equipment becomes considerably simpler in a continuous process.

LABOR SAVINGS

In the present batch operations a highly skilled sugar boiler is required for each pan in addition to centrifuge operators and dryer operator. Although the number varies from plant to plant, it is estimated that about six operators are required for the centrifugation, drying, and crystallization steps. In a continuous system this number can be reduced to two.

OPERATIONAL ADVANTAGES

The product from a continuous crystallization system has more simple grain formation and lacks agglomerate and dendritic crystals, thereby resulting in a product which has a more pleasing appearance. The higher volumetric weight will give some savings in packaging material.

The particle size control in a continuous system is considerably simpler than in a batch system, where the results depend to such a large degree on the skill of the operator. The low slurry density used in continuous processing lends itself more favorably to pumping and handling than the extremely heavy magmas which are typical of batch operation.

Instrumentation for the control of a continuous plant is inherently more simple and easier to maintain than the cyclic controls required by batch operation. The only basic instrumentation required for control is an absolute pressure controller, a steam flow recorder-controller, a supersaturation indicator, and a liquid level control.

Since the large feed valves and large foot valves along with their associated pumps and piping which are required to fill and empty the batch machines quickly can be eliminated in a continuous process and replaced with relatively inexpensive air operated control valves, the cost of the instrumentation and its maintenance will be significantly lower.

SUMMARY

Based on the test program covered in this paper, it appears that the growth of sugar follows the McCabe delta L law and gives behaviors expected and predicted from the work of Randolph and Larson and Larson's coworkers at the Iowa State University. Growth rates of approximately 0.007 to 0.09 mm./hr. were measured at supersaturation coefficients of 1.2 to 1.8. Within this range the growth rate was directly proportional to supersaturation. These values are in the same order of magnitude as values reported in the literature, although somewhat lower. The use of continuous crystallization appears to be feasible, and the continuous crystallizer described may be controlled by regulating the amount of mechanical stimulation and, hence, the nucleation rate. The physical properties of sugar produced in continuous crystallization machines operating at supersaturation coefficients of around 1.4 and growth rates of approximately 0.01 to 0.05 appear to be superior in terms of purity, ash, and color than the conventional product made in batch crystallizers. Process analysis indicates that the utility consumption for a continuous system should be substantially less than for most present batch type of crystallizers. Based on the information and work done and described in this report, the use of continuous sugar crystallization appears attractive, possible, and economically justified.

NOTATION

$$cv = \text{coefficient of variation} = 100 \left[\frac{P.D._{16\%} - P.D._{84\%}}{2\,(P.D._{50\%})} \right]$$

n° = population density, No./mm. (per liter of slurry)

r = growth rate, mm./hr.

s = supersaturation coefficient,

$$\frac{\text{sugar/water (in solution)}}{\text{sugar/water (at saturation)}}$$

LITERATURE CITED

1. Honig, P., "Principles of Sugar Technology," Vol. 2, p. 448, Elsevier, New York (1959).
2. Randolph, A. D., and M. A. Larson, *AIChE J.*, **8**, No. 5, 639 (Nov., 1962).
3. Honig, P., "Principle of Sugar Technology," Vol. 2, p. 156, Elsevier, New York (1959).
4. Van Hook, A., "Crystallization, Theory and Practice," p. 244, Reinhold, New York (1961).
5. Chapman, F. M., private communication dated Jan. 16, 1961, from the data of Golovin.
6. Gillette, E., "Low Grade Sugar Crystallization," C & H Sugar Refining Corporation, Crockett, Calif. (1948).
7. Randolph, A. D., *AIChE J.*, **11**, 424 (1965).

COMMERCIAL UREA CRYSTALLIZATION

The crystallization of urea by a continuous method is described. The influence of biuret concentration is discussed, and various methods of correlating the size distribution data are illustrated. The results show some inconsistency by using some of the published analysis techniques. It was concluded that in this instance the process lies outside the scope of the analysis used.

Richard C. Bennett
and
Maarten Van Buren

Within the last 3 yr. there has been a very rapid increase in the usage of urea as a high grade fertilizer, animal feed supplement, and raw material for certain resins. The low-cost production of urea in large quantities has been widely studied by chemical engineers, and a number of processes have been developed, both in this country and abroad, which reflect considerable sophistication in engineering technology and in the use of specialized alloys to meet the corrosion problems encountered in the reactions required to produce urea.

At least four (*1* to *5*) different types of urea processes are currently being marketed, and these have been amply described in the literature. The bulk of the urea production in the United States is in the form of spherical prills produced in one of two ways. By starting with a reactor effluent which contains about 70 to 80% urea by weight plus 0.3 to 0.5% biuret, the solution is pumped to an evaporator which has either one or two stages. The evaporator raises the concentration of the urea to about 99.7% by weight. During this concentration there is an inevitable decomposition of the urea into biuret and water as the result of the residence time and temperature during the evaporation cycle. Particularly damaging is the high temperature at the highest concentration. Prills made in this way contain around 0.4 to 1% biuret.

Although this product has been suitable for many markets in the United States and abroad, there is an increasing demand for crystalline urea and prills with very low biuret which can be used for cattle feed supplements and plastics. For such uses, the biuret in the final product should not exceed 0.2 to 0.3%. Since the amount of biuret in the final product is so critical, it is not surprising that other methods of separating the urea from the solution containing biuret and other impurities should come into existence. Most plants wishing to produce a low biuret product employ some type of crystallization to separate a very pure urea crystal from the solution produced by the reaction system and then melt this crystal with a very short residence time before prilling the final product. Alternatively, the washed and dried crystalline material can be also marketed for certain specific uses such as the production of urea formaldehyde resins.

Swenson, Division of Whiting Corporation, Harvey, Illinois.

THE CRYSTALLIZATION PROCESS

The production of crystalline urea and the techniques for separating it from a solution which contains biuret can best be explained by referring to a phase diagram of the urea, biuret, and water systems shown in Figure 1

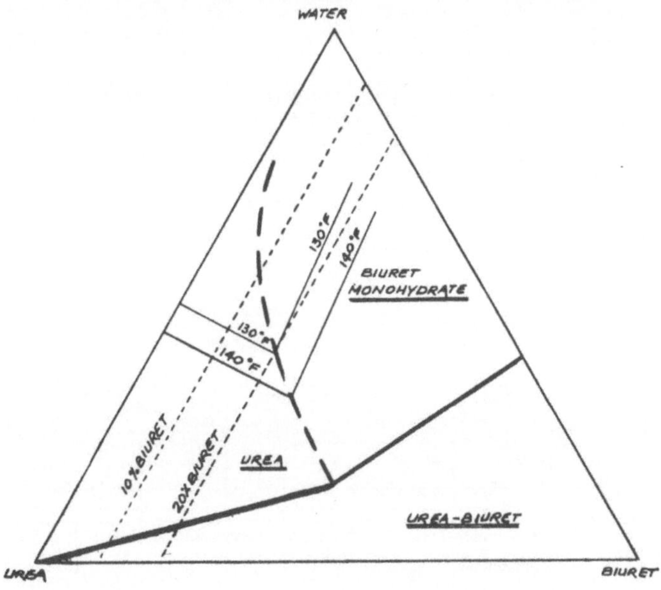

Fig. 1. Urea-biuret-water phase diagram.

(6). Solid phase urea will exist in the lower left-hand corner of the diagram to the left of the dotted line, and for solution compositions represented by points on the dotted line a mixed crystal will be produced consisting of biuret, menohydrate and urea. According to the diagram, which is borne out in practice, solutions of up to 20% biuret may be used when crystallizing urea providing the temperature exceeds 130°F. In practice, it is generally desirable not to go above approximately 10% biuret in crystallizer solutions so as to make washing on the centrifuge less critical. Urea solutions have boiling point elevations which are about 20° to 25°F., depending on the temperature, and the viscosity of urea solutions at about 130°F. is approximately 1.14 centipoise. Since the density of a urea crystal is only 1.335 vs. a solution specific gravity of about 1.15 to 1.17, the settling rate of the crystals, irrespective of size, is not very high. Urea crystals are white rhombohedrons with a length to width ratio of about 2.5 to 1.0, depending on the biuret concentration in the mother liquor and the rate of growth.

Since urea has normal solubility and can be easily produced in highly concentrated solutions, it is not surprising that most crystallization takes place in equipment which utilizes sensible heat plus the heat of crystallization to produce most of the required evaporation. Solutions below about 70% urea require evaporative crystallizers, however. Swenson has produced large-scale crystallizers for urea in both the forced circulation and draft tube baffle (DTB) configurations. Preliminary work on both of these designs was done on pilot plant scale in equipment with an active volume of about 200 gal. In addition, work has been done in relatively small glassware having an active volume of about 2 liters.

The forced circulation crystallizer mentioned in the previous paragraph is shown in Figure 2 and consists of a crystallizer body with an external recirculating loop in which the urea slurry is circulated by an axial flow pump. If the unit requires substantial quantities of evaporation, then a heating element of the conventional tube and shell type is placed in this circulation loop. In all cases, the speed of the recirculation pump, its type and design, as well as the retention in the body are important considerations so that the final product specifications can be satisfied.

Shown in Figure 3 is a crystallizer of the draft tube baffle type which has been discussed a number of times in the literature (7 to 9). The main distinction between this machine and the forced circulation machine is in the use of an internal settling zone which permits the separation of fine crystals which are then destroyed so that the production is forced to take place on a fewer number of seed crystals, thereby increasing the average particle size. In this design, a large internal propeller imparting characteristically low mechanical stimulation to the slurry is employed to give the needed internal recirculation to maintain the suspension and minimize the supersaturation created by vapor release at the boiling surface. The settling zone must be designed so that a separation of product size crystals from the fine crystals which are to be subsequently destroyed can be made without imposing too great a burden of recrystallization due to the quantity of fines destroyed. Such a machine

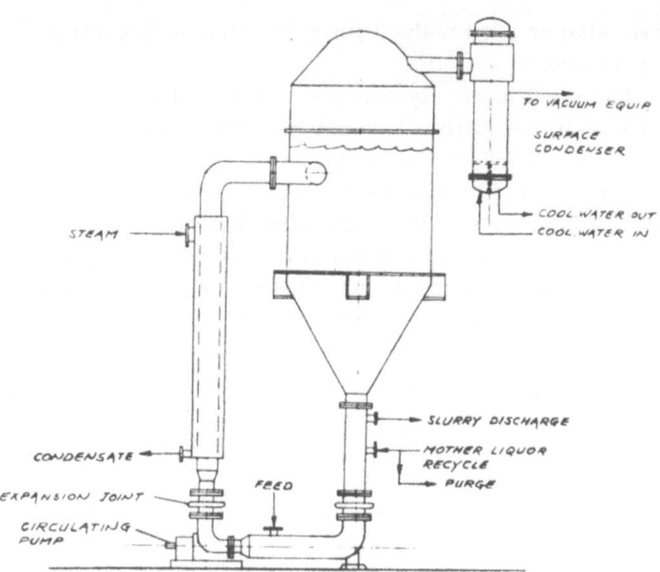

Fig. 2. Swenson forced circulation urea crystallizer.

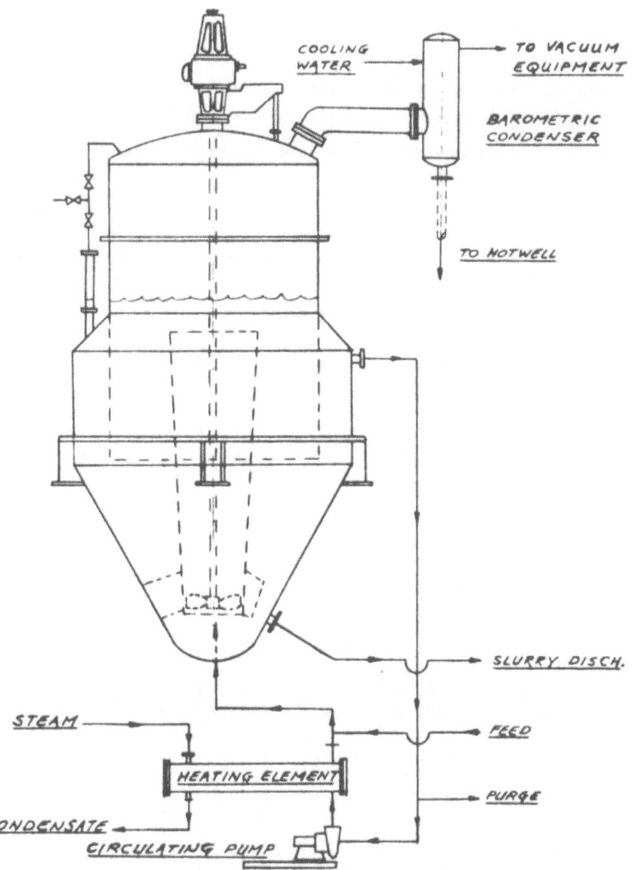

Fig. 3. Swenson DTB urea crystallizer.

may also be used without fines destruction on cooling applications.

The yield in the crystallizer is achieved by the change in solubility which occurs between the feed temperature and the mother liquor temperature and the vaporization of water. Heat to vaporize this water comes from the sensible heat of the feed and the heat of crystallization of urea which is relatively high, approximately 104 B.t.u./lb. In some cases, owing to washing on the centrifuge or use of a more dilute feed, it is necessary to supply additional heat, and this can be done through the heating element described which is located either in the slurry recirculation loop in the forced circulation machine or in the fines destruction loop in the DTB design. The heat may come from process steam or from heat generated in the reaction system itself.

RESULTS

The crystalline product produced by both types of urea crystallizers is ordinarily stable under steady operating conditions and is a crystalline material averaging between 20 and 50 U.S. mesh. Summarized in Table 1 are the growth rates, residence times, and screen analyses for various types of machines operating under different conditions. The growth rates have been calculated in the manner outlined by Randolph and Larson (*10, 11*) and in subsequent papers by Larson and his co-workers at the Iowa State University. In addition to the growth rate (which is the diametrical growth rate), there is also the logarithm of the nuclei population density. This nuclei population density has been calculated on the basis of the number of crystals in a liter of crystallizer volume. A typical growth rate plot is shown in Figure 4, where the natural logarithm of the population density is plotted vs. crystal length. As can be seen, this plot is a straight line. This straight line is expected if the basic criteria used in the derivation of the formulas were correct. These criteria are: that the suspension is thoroughly mixed; that a mixed product is removed from the crystallizer, that is, fine crystals, product size crystals, etc., without any classification; that the system is operating under steady state conditions; and that growth proceeds in accordance with the McCabe delta L law, which means that the advance of a crystal face is linear with respect to time. For the samples shown in Table 1, the DTB crystallizer was operated without fines removal so that the conditions underlying calculation of the growth rates, according to this technique would be fulfilled. By assuming a growth rate, a nuclei population, and a retention time, the projected particle size may be computed. This has been done for one example and is shown in the Appendix. It is seen that it bears a close resemblance to the actual measured screen analysis.

By assuming that the nuclei population density can be represented by a simple power relation as outlined in the paper by Larson, Timm, and Wolff (*11*) as shown in their Equation (20), the nuclei population density is

$$n^{\circ} = K_m M^j r^{i-1}$$

By graphing the logarithm of the nuclei population density vs. the logarithm of the growth rate, a straight line should result whose slope is equal to $(i - 1)$. This has been done and is shown in Figure 5 for samples from the DTB crystallizers having approximately the same slurry density. Since the i value would be expected to be the

TABLE 1

Sample No.	Growth Rate MM/HR.	Growth Rate ln.r	Nuclei Population ln.n°	Density gpl	Residence Time Hrs. T	Tyler Screen Sizes 14	20	28	35	48	65	100	Type Mach.
113	0.0485	-3.03	16.3	510	5.90	36.0	62.0	80.5	92.5	98.5	100		DTB - Large
116	0.0800	-2.53	16.3	317	3.12	26.2	51.5	76.2	94.0	99.5	100		"
117	0.0495	-3.01	17.5	404	3.97	19.0	38.5	63.5	81.5	98.0	100		"
118	0.0445	-3.11	17.8	442	4.00	13.0	31.0	54.5	75.0	92.5	100		"
121	0.0690	-2.67	16.5	400	3.92	3.5	44.5	73.5	91.5	98.5	100		"
135	0.0259	-3.65	17.9	560	6.70	-	45.0	65.0	84.0	95.0	100		"
136	0.0240	-3.73	18.2	540	6.70	2.0	38.0	63.0	85.5	96.5	100		"
124	0.1500	-1.90	15.8	435	2.30	8.0	60.0	81.0	94.0	98.0	100		DTB - Small
125	0.1180	-2.13	16.5	440	2.30	17.0	54.0	83.0	96.0	99.0	100		"
129	0.0429	-3.15	17.7	475	4.58	-	26.0	62.0	83.0	95.0	100		FC - Small - (A)
130	0.0350	-3.35	18.4	296	3.90	-	8.0	41.0	72.0	92.0	100		"
131	0.0234	-3.76	19.6	285	4.40	-	3.0	30.0	63.0	87.0	100		"
132	0.0236	-3.75	18.8	392	5.35	-	-	24.0	64.0	88.0	100		"
133	0.0140	-4.27	20.3	286	6.60	-	-	10.0	43.0	75.0	100		"
191	0.0330	-3.41	18.81	450	4.00	-	4.6	45.1	75.8	88.4	96.8	100	FC - Large - (B)
192	0.0338	-3.43	19.71	450	3.38	-	4.4	18.8	43.0	74.6	90.1	97.5	"
193	0.0317	-3.45	18.70	450	4.57	-	13.6	59.1	79.8	84.4	96.3	97.9	"
194	0.0200	-3.91	20.51	450	4.57	-	3.1	23.7	48.2	68.6	83.4	91.9	
197	0.0624	-2.77	18.1	405	2.50	1.7	15.1	50.6	79.2	91.7	98.0	99.8	FC - Large - (C)
196	0.0580	-2.84	18.5	400	2.50	4.7	10.5	33.8	69.3	88.5	97.7	99.9	"
195	0.0413	-3.18	20.0	395	2.50	4.3	8.2	18.2	44.4	74.4	90.5	98.3	"

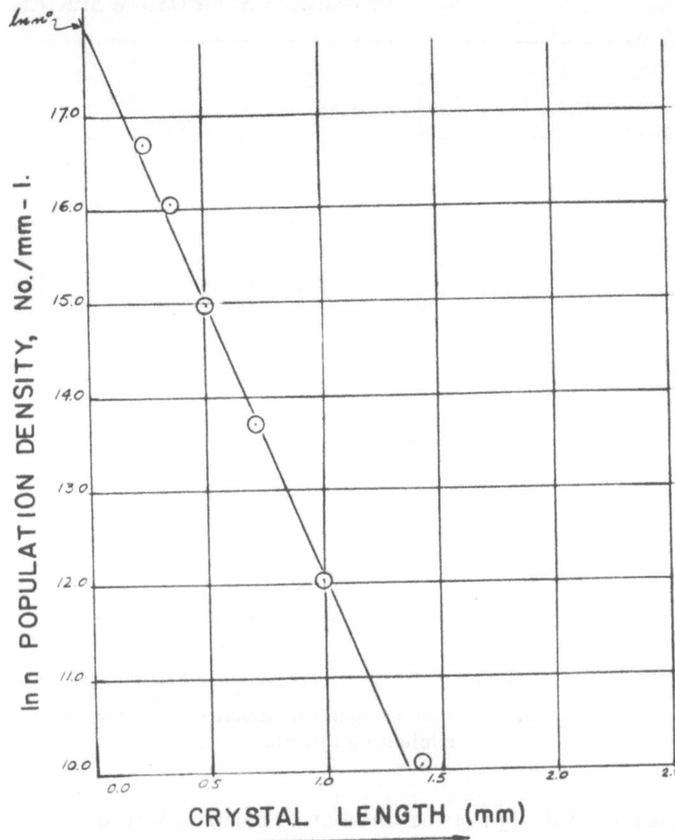

Fig. 4. Plot of the log of the population density vs. crystal length.

same for all slurry densities, a line of constant slope has been used to connect the various points having similar densities. The calculated slope for this line is −1.66. The data from this graph represents samples taken from two machines having greatly different volumes. Points 125 and 124 were from the laboratory machine with an active volume of 200 gal. The remaining points were taken from a large-scale machine with an active volume of about 25,000 gal. Therefore, the scale-up between these two lines represents a ratio of about 125:1.

In Figure 6 is a similar plot showing the logarithm of the population density plot vs. the logarithm of the growth rate for various sizes of forced circulation machines. Points 129 to 133 are from the laboratory forced circulation machine with an active volume of approximately 175 gal. The remaining points are taken from large-scale forced circulation machines having active volumes of about 11 to 12,000 gal.

The laboratory machine shows lines whose slopes are equivalent to an i value of −1.66. In these examples there is little difference between the lines representing a density of about 286 and 475 g./liter. In addition, the lines from the two larger machines at approximately 400 to 500 g./liter slurry density are represented by slopes of $i = -2.20$ and $i = -3.45$.

It would not be expected that i would have a negative exponent, since there is no physical reason to expect

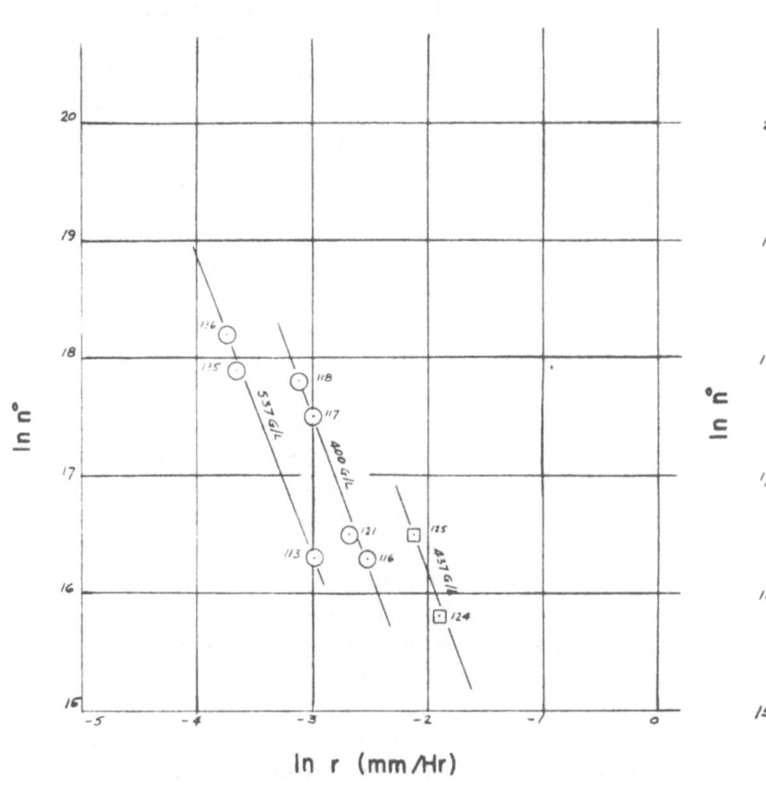

Fig. 5. Plot of the log of the nuclei population vs. the log of the growth rate in draft tube crystallizers.

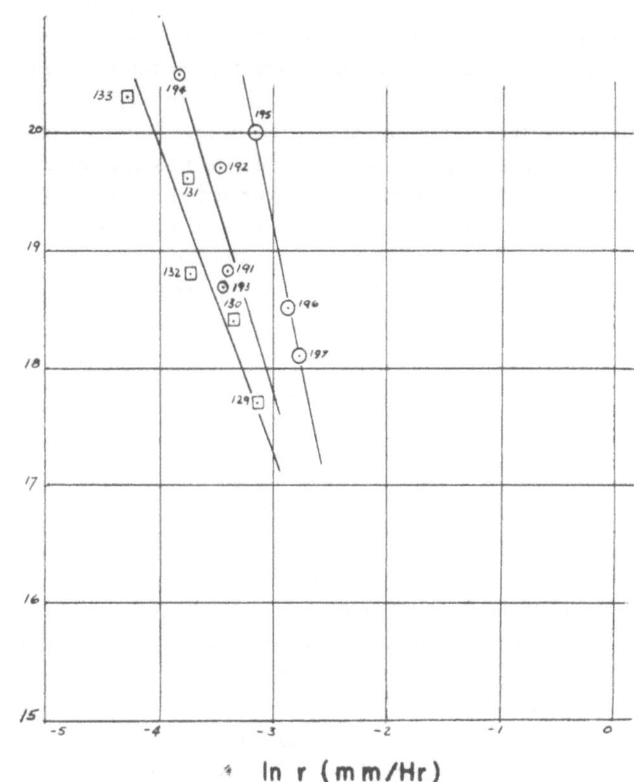

Fig. 6. Plot of the log of the nuclei population vs. the log of the growth rate in forced circulation crystallizers.

that the nucleation rate would ever decrease with increasing growth rate. In most systems reported in the literature the value of i is either 1 or a power higher than 1. The answer must, therefore, lie outside the scope of the derivation proposed by Larson, et al. (*10*).

In the forced circulation system shown in Figure 6, the lines of the highest negative slope are from systems with the highest level of mechanical stimulation. There is also some tendency evident from examining the data to note that the larger products are produced at shorter residence time. Both of these facts would be expected if mechanical stimilus and attrition were controlling factors in determining the nucleation rate. In any real system the nucleation is inevitably the sum of that created from supersaturation and from reduction of larger particles due to attrition. The nucleation from supersaturation is a complex function in itself, being caused by the instability of the driving force, mechanical effects of the system on this instability, and the modification of these effects due to impurities accidentally or purposely present. Any real crystallizer produces a product which is the combination of all of these influences, and it has long been recognized that increasing retention times do not always result in larger product crystals because the influence of longer periods for growth can be more than offset by the increased nucleation due to attrition and mechanical stimulation which occur during this correspondingly longer retention period.

Early work by Young and his co-workers (*12*) in 1911 demonstrated that mechanical stimulus through both percussion and spinning of stirrers could promote nucleation. Van Hook (*13*) demonstrated that ultrasonic irradiation could also produce significant increases in nucleation, and the author has indicated in previous papers (*7*, *8*) that mechanical stimulus, defined as the total influence of the mechanical and hydraulic regime on the rate of nucleation, were important factors in commercial crystallizer design. In the data incorporated as Table 1 of this paper, points 191 through 194 are from a large machine embracing one hydraulic regime, and points 195 through 197 are from a second large forced circulation machine employing a different hydraulic regime. Although at present there is no recognized dimensionless group or characteristic parameter that can be used to characterize this nucleating stimulus, the data for the three forced circulation machines can be compared on a relative basis by selecting a common turbulent element in comparing the systems with each other. This has been done and has been shown in Figure 7. On the ordinate axis the natural log of nucleii population density is plotted for values selected from Figure 6 at constant growth rate. As can be seen in Figure 7, there is a relatively good correlation between the data from large and small scale equipment when plotted on this basis.

Experience has shown that it is essential in large-scale equipment design to pay careful attention to the

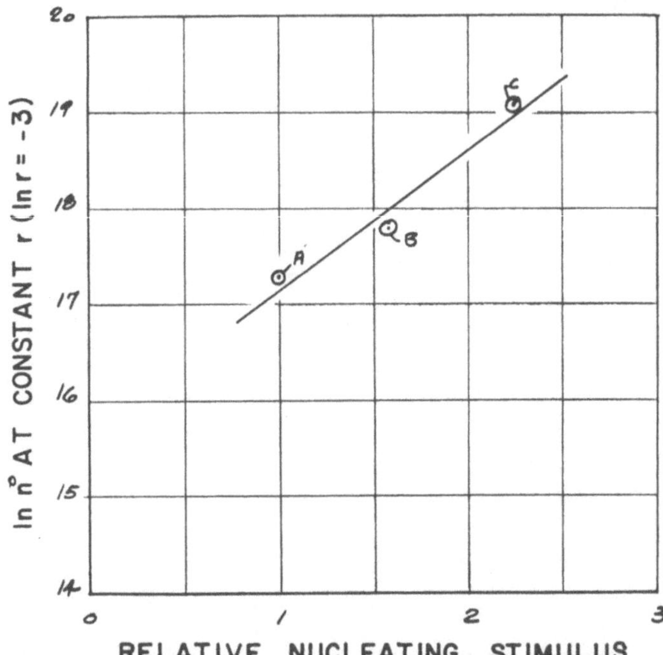

Fig. 7. Change in nuclei population density with change in nucleating stimulus.

speed and design of the circulating pump as well as to the inlet conditions at the body so as to minimize the influence of this circuit on nucleation in the system and to achieve similitude in scale-up.

By using Equation (20) in Larson, Timm, and Wolff's paper, the large draft tube system shown in Figure 5 was compared at equal growth rates, and the power of j was determined to be -3.87. This indicates that as the slurry density is increased, the nuclei population density decreases. This would be expected, since with most crystallization systems increasing the slurry density results in an increase in particle size, that is, less nucleation. Unfortunately, there was insufficient data from the same forced circulation machine operating at various densities to get a correlation on the value of j for the forced circulation system.

Based on the data presented above, it would appear that large-scale urea crystallizers may confidently be designed and the screen analyses of the product accurately predicted for systems of greatly different sizes. The crystals tend to be blocky when the biuret level in the mother liquor is in the range of 3 to 10%. It currently appears that about the best ratio is in the neighborhood of about 8% biuret. Apparently biuret in the system influences the L/D ratio of the crystals, with small values of biuret causing needlelike crystals to be formed, whereas values in the range of 5 to 8% result in the production of chunky crystals.

Based on measurements made by repeated washing of samples of the crystals from the crystallizer, it would appear that the occluded biuret formed when crystallizing urea from solutions containing about 5 to 8% biuret would be in the order of 0.01% to 0.02%. Such a product may be readily centrifuged with proper washing to a biuret content of approximately 0.05 to 0.1% biuret.

CONCLUSIONS

It would appear that the crystallization of urea from solutions containing appreciable quantities of impurities, such as biuret, may be readily performed to yield a crystal containing about 99.98% urea. Such crystals are well defined and relatively easy to wash and may be produced in tonnage quantities with techniques and equipment that are both predictable in operating characteristics and economically feasible.

LITERATURE CITED

1. Weyermuller, Gordon, W. S. Schellentrager, and L. T. Marks, *Chem. Process.*, 19 (Sept. 24, 1962).
2. Reynolds and Tremarke, *Hydrocarbon Process. Petrol. Refiner*, 109 (Dec., 1962).
3. Guccione, Eugene, *Chem. Eng.* (Jan. 18, 1965).
4. Taggrasco, Ronald, *ibid.* (Aug. 1, 1966).
5. Guccione, Eugene, *ibid.* (Sept. 26, 1966).
6. de Malde', *La Chemical E L'Industria*, **38**, 572 (1956).
7. Newman, H. H., and R. C. Bennett, *Chem. Eng. Progr.*, **55**, No. 3 (Mar., 1959).
8. Bennett, R. C., *ibid.*, **58**, No. 9 (Sept., 1962).
9. Caldwell, *Ind. Eng. Chem.*, **53**, 115–8 (Feb., 1961).
10. Randolph, A. D., and M. A. Larson, *AIChE J.*, **8**, 639 (1962).
11. Larson, M. A., D. C. Timm, and P. R. Wolff, *ibid.*, **14**, No. 3, 448 (May, 1968).
12. Buckley, H. E., "Crystal Growth," p. 19, Wiley, New York (1951).
13. Van Hook, Andrew, "Crystallization," p. 206, Reinhold, New York (1961).

APPENDIX: CALCULATION OF SCREEN SIZE DISTRIBUTION FROM GROWTH RATE DATA

Sample 118 $r = 0.0445$ mm./hr.

$T = 4.00$ hr.

Slurry Density = 442 g.p.l.

$n^\circ = e^{17.8}$, $\ln n^\circ = 17.8$, No./mm. – liter

$$\ln n = \frac{-L_{avg}}{rT} + \ln n^\circ$$

14 Mesh Fraction: $\ln n = \dfrac{-1.409}{(.0445)(4)} + 17.8 = 9.9$

$$n = 2(10^4) \text{ or } e^{9.9}$$

$$\text{weight/liter} = \left(n\,\frac{\text{No.}}{\text{mm.}}\right)(\Delta L\text{mm.})\left(\frac{\text{g.}}{\text{xtal}}\right)$$

$$= 2(10^4)(0.484)(0.00373)$$

$$= 36 \text{ g.}$$

20 Mesh Fraction: $\ln n = \dfrac{-1.00}{(.178)} + 17.8 = 12.2$

$$\text{wt.} = e^{12.2}(0.335)(0.00133) = 89 \text{ g.}$$

28 Mesh Fraction: $\ln n = \dfrac{-0.711}{(.178)} + 17.8 = 13.8$

$$\text{wt.} = e^{13.8}(0.244)(0.000479) = 116 \text{ g.}$$

35 Mesh Fraction: $\ln n = \dfrac{-.503}{(.178)} + 17.8 = 14.98$

$$\text{wt.} = e^{14.98}(0.172)(0.0001699) = 93.8 \text{ g.}$$

48 Mesh Fraction: $\ln n = \dfrac{-0.356}{0.178} + 17.8 = 15.8$

$$\text{wt.} = e^{15.8}(0.122)(0.00006023) = 53.5 \text{ g.}$$

65 Mesh Fraction: $\ln n = \dfrac{-0.251}{0.178} + 17.8 = 16.4$

$$\text{wt.} = e^{16.4}(0.087)(0.00002111) = 23.9 \text{ g.}$$

Size	L_{avg}, mm.	ΔL, mm.	g. Crystal	Weight liter	Wt. %	Actual wt. %
14	1.409	0.484	0.00373	36.0	8.8	13.1
20	1.000	0.335	0.00133	89.0	21.6	18.1
28	0.711	0.244	0.000479	116.0	28.2	23.6
35	0.503	0.172	0.0001699	93.8	22.8	20.1
48	0.356	0.122	0.00006023	53.5	13.0	17.6
65	0.251	0.087	0.00002111	23.9	5.6	7.5

DESIGN MODELS FOR CONTINUOUS CRYSTALLIZERS WITH DOUBLE DRAWOFF

Hugh M. Hulburt
and
Denis G. Stefango

A particle balance model is developed for a well-stirred continuous crystallizer with two discharge streams, one containing only fines and mother liquor and one containing the average contents of the crystallizer. Steady state equations for the operation of such a system are developed and examined for conditions leading to cyclic variation of crystal size. Computational procedures are used which can be extended to more detailed and more realistic models.

Models for the performance of continuous crystallizers have been proposed and examined by Saeman (6), Randolph and Larson (5), Murray and Larson (3), Randolph (4), and Sherwin, Shinnar, and Katz (7) and have been set in the context of the general population balance treatment by Hulburt and Katz (2). Let $f(r, x; t)$ be the number of crystals of size r per unit volume of crystallizer per unit size at a given position x and moment t. The balance equation for f then reads

$$\frac{\partial f}{\partial t} + \frac{\partial}{\partial x}(\dot{x}f) + \frac{\partial}{\partial r}(\dot{r}f) = h(r, x; t) \quad (1)$$

The life history of a crystal is defined by its size $r(t)$ and its location in the crystallizer $x(t)$. It can be represented as a trajectory in x, r space, the phase space of the process. At some point (x_0, r_0) the crystal is born by nucleation or by injection from outside, and at some point (x_1, r_1) it leaves the crystallizer. In the absence of agglomeration, each crystal follows a continuous trajectory from (x_0, r_0) to (x_1, r_1). Then $f(x, r; t)$ is the density of phase points about (x, r) at the moment t. The left side of Equation (1) describes the change in f due to convection along the continuous trajectories, while the function $h(x, r; t)$ is the source term which gives the net rate of increase of new trajectories in the volume element about (x, r).

In the interior of a continuous crystallizer, $h(x, r; t)$ will be the nucleation rate. Since the flow patterns in a crystallizer are very complicated and not well characterized, it is usual to approximate them by considering the crystallizer to be a set of macroscopic connected regions with sensibly uniform flows and compositions in each region. Appropriate material balance equations connect the flows and composition of one region to the next. We thus integrate Equation (1) over the volume of each region:

$$\bar{f} = \int_V f(x, r; t)\, dx \quad (2)$$

We note that Gauss' theorem states

$$\int_V \nabla \cdot \varphi \, dx = \int_s \varphi_N \cdot d\sigma \quad (3)$$

where φ is the outward vector flux of a property such as $\dot{x}f$. Hence

$$\int_V \nabla \cdot (\dot{x}f)\, dx = \int_s (\dot{x}f)_N \cdot d\sigma = \text{outflow} - \text{inflow} \quad (4)$$

and

$$\frac{\partial \bar{f}}{\partial t} + \frac{\partial}{\partial r}(\dot{r}\bar{f}) = \bar{h}(r; t) + \text{(inflow of crystals of size } r - \text{outflow of crystals of size } r) \quad (5)$$

The crystal growth rate is expressed in $\dot{r}$ when this is

Northwestern University, Evanston, Illinois. Denis G. Stefango is with the Horton Process Division, Chicago Bridge and Iron Company, Chicago, Illinois.

specified as a function of ambient conditions and, possibly, crystal size.

Equation (5) is applicable to the entire body of a well-stirred crystallizer, in which it is now supposed that the nucleation and growth rates can be expressed in terms of the average conditions of temperature and supersaturation in the body. If there are local regions where these properties differ greatly from the average, they are isolated as separate stages, each treatable by Equation (5).

In the strictly countercurrent classifying crystallizer, Equation (5) is not useful, but $\dot{x}$ is now well defined by the plug flow model adopted for the flow pattern. Inflow and outflow now appear as boundary conditions in Equation (1) at the inlet and outlet values of x.

For the simplest case of a single well-stirred crystallizer with a single inlet and a single outlet with no internal classification, Equation (5) has been solved in terms of arbitrary nucleation and growth kinetics. The stability of the steady state solutions has also been examined by Sherwin and Katz (*7*). The useful design information is contained in the moments of $f(r, t)$, namely

$$\mu_0 = \int_0^\infty f(r, t)\, dr \tag{6}$$

$$\mu_1 = \int_0^\infty r\, f(r, t)\, dr \tag{7}$$

$$\mu_2 = \int_0^\infty r^2\, f(r, t)\, dr \tag{8}$$

$$\mu_3 = \int_0^\infty r^3\, f(r, t)\, dr \tag{9}$$

These are proportional to the average crystal number density, number average size, area, and mass, respectively. Note that the usual size distributions are reported in mass of crystals of size r. This distribution f_w is given by

$$f_w(r, t)\, dr = \Phi\, r^3 f(r, t)\, dr \tag{10}$$

where Φ is the shape-density factor such that $\Phi\, r^3$ is the mass of a particle of size r. Thus, when Φ is independent of r, the mass average size is

$$\overline{r_w} = \int_0^\infty r\, f_w\, dr = \int_0^\infty \Phi\, r^4 f(r)\, dr = \Phi\, \mu_4 \tag{11}$$

The mass average surface area is

$$\overline{r_w^2} = \int_0^\infty \Phi\, r^5 f\, dr - \Phi\, \mu_5 \tag{12}$$

Equation (5) can be used to derive a set of five equations for the moments μ_0, μ_1, μ_2, and μ_3 and the mother liquor concentration c. This set is closed and can be solved without finding $\overline{f}(r)$ explicitly. However, when there are multiple inlet and outlet points, or when there is product classification with recycle, either internally or externally, the moment equations no longer close in general, and Equation (5) must be added explicitly to the set.

Hulburt and Akiyama (*1*), acting on a suggestion of Hulburt and Katz (*2*), have shown how rapidly converging iterative solutions can be obtained in this more general case by choosing an approximate form for $\overline{f}(r)$ with parameters which can be adjusted at each stage to reproduce the leading moments. Equation (5) is then bypassed and the approximate $\overline{f}$ used to calculate the extra terms in the moment equations.

In this paper, this procedure will be used to work out design equations for a well-stirred crystallizer with internal classifier and separate drawoff of mother liquor containing entrained fines from heavy product in mother liquor. This double drawoff permits independent control of magma density. The mother liquor plus fines may be recycled either with or without redissolving fines. In this study, only the case of clear mother liquor drawoff was considered in detail, although provision was made for independent control of feed stream crystal size distribution. For simplicity, it was assumed that the internal classifier simply split the magma into a mother liquor stream discharged with entrained fines through one line and a heavy product consisting of the remaining crystals discharged with mother liquor through another line. The size distribution in each exit line was thus simply related to the size distribution in the body as a whole. This is obviously an oversimplification adopted for computational convenience. More realistic models could be treated analogously to this case but would require heavier, though manageable, computation.

A schematic flow sheet is shown in Figure 1. The

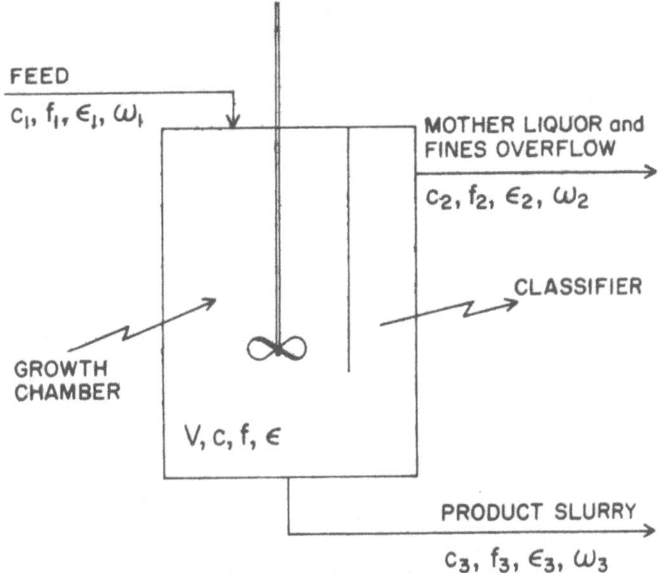

Fig. 1. Schematic diagram of classifying crystallizer.

moment equations are

$$\frac{d\mu_0}{dt} = (1 - k\,\mu_3)\,B(c) - \frac{\omega_3}{V}\mu_0 - \frac{\mu_2}{V}\nu_0 + \frac{\omega_1}{V}\xi_0 \quad (13)$$

$$\frac{d\mu_1}{dt} = G(c)\,\mu_0 + (1 - k\,\mu_3)\,B(c)\,r_0 - \frac{\omega_3}{V}\mu_1 - \frac{\omega_2}{V}\nu_1 + \frac{\omega_1}{V}\xi_1 \quad (14)$$

$$\frac{d\mu_2}{dt} = 2\,G(c)\,\mu_1 + (1 - k\,\mu_3)\,B(c)\,r_0^2 - \frac{\omega_3}{V}\mu_2 - \frac{\omega_2}{V}\nu_2 + \frac{\omega_1}{V}\xi_2 \quad (15)$$

$$\frac{d\mu_3}{dt} = 3\,G(c)\,\mu_2 + (1 - k\,\mu_3)\,B(c)\,r_0^3 - \frac{\omega_3}{V}\mu_3 - \frac{\omega_2}{V}\nu_3 + \frac{\omega_1}{V}\xi_3 \quad (16)$$

$$(1 - k\,\mu_3)\frac{dc}{dt} = \frac{\omega_1}{V}(c_1 - c)\,\varepsilon_1 - (\rho - c) \times [3\,G(c)\,k\,\mu_2 + (1 - k\,\mu_3)\,B(c)\,k\,r_0^3] \quad (17)$$

Equation (5) takes the form

$$\frac{\partial f}{\partial t} + G(c)\frac{\partial(\Phi f)}{\partial r} = \varepsilon\,B(c)\,\delta(r - r_0) - \frac{\omega_3}{V}f - \frac{\omega_2}{V}f_2 + \frac{\omega_1}{V}f_1 \quad (18)$$

By material balance, the equation for the mother liquor concentration is

$$\mu_2 = \left\{\left(\frac{\varepsilon B}{\overline{G}}\right)\left(\frac{V\overline{G}}{\overline{\omega}_1}\right)^3[(d\,r_0'^2 + 2\,d^2 r_0' + 2\,d^3) - e^{-\theta_1}(d\,r_c'^2 + 2\,d^2 r_c' + 2\,d^3)]\right\}_A + \left\{\left(\frac{\varepsilon B}{\overline{G}}\right)\left(\frac{V\overline{G}}{\overline{\omega}_1}\right)^3\left[e^{-\theta_1}\left(\frac{r_c'^2}{x} + \frac{2\,r_c'}{x^2} + \frac{2}{x^3}\right)\right]\right\}_B \quad (22)$$

$$\mu_3 = \left\{\left(\frac{\varepsilon B}{\overline{G}}\right)\left(\frac{V\overline{G}}{\overline{\omega}_1}\right)^4[(d\,r_0'^3 + 3\,d^2 r_0'^2 + 6\,r_0' d^3 + 6\,d^4) - e^{-\theta_1}(d\,r_c'^3 + 3\,d^2 r_c'^2 + 6\,d^3 r_c' + 6\,d^4)]\right\}_A + \left\{\left(\frac{\varepsilon B}{\overline{G}}\right)\left(\frac{V\overline{G}}{\overline{\omega}_1}\right)^4\left[e^{-\theta_1}\left(\frac{r_c'^3}{x} + \frac{3\,r_c'^2}{x^2} + \frac{6\,r_c'}{x^3} + \frac{6}{x^4}\right)\right]\right\}_B \quad (23)$$

$$\varepsilon\frac{\partial c}{\partial t} = \frac{\omega_1}{V}(c_1 - c)\,\varepsilon_1 - (\rho - c)(\sigma\,G + \varepsilon\,B\,k\,r_0^3) \quad (18a)$$

The operating variables may be expressed dimensionlessly for greatest generality. The characteristic times of the process are magma residence time, nucleation half-life, and growth half-life. The split between overhead and product is the fourth definitive operating variable. Feed composition and feed crystal size distribution enter the terms defining the characteristic times. Thus, we write $r' = (\overline{\omega}_1 r)/(V\overline{G})$ for the dimensionless crystal size in terms of the size $V\overline{G}/\overline{\omega}_1$ of the crystal in steady state, whose age is one residence time, $V/\overline{\omega}_1$. Similarly, the nucleus size is $r_0' = \overline{\omega}_1 r_0/V\overline{G}$, and the largest crystal in the overflow has size $r_C' = \overline{\omega}_1 r_C/V\overline{G}$. The number density $\overline{f}$ is normalized in terms of the number of nuclei formed per unit time in steady state in the mother liquor $\varepsilon\,\overline{B}/\overline{G}$ to give $f' = \overline{f}/(\varepsilon\,\overline{B}/\overline{G})$ as the normalized number density. In steady state, with no nuclei entering with the feed

$$\overline{f}' = \overline{f}/(\varepsilon\,\overline{B}/\overline{G}) = \exp\left\{-\left[\frac{1 - x(1 - \varepsilon)}{\overline{\varepsilon}}\right](r' - r_0') + U(r_c')\frac{(1 - x)}{\overline{\varepsilon}}(r' - r_c')\right\} \quad (19)$$

where $U(r_c')$ is the unit step function at $r' = r_c'$ and $x = \overline{\omega}_3/\overline{\omega}_1$ is the ratio of product slurry volume flow to feed volume flow.

The steady state moments are

$$\overline{\mu}_0 = \left\{\left(\frac{\varepsilon B}{\overline{G}}\right)\left(\frac{V\overline{G}}{\overline{\omega}_1}\right)d[1 - e^{-\theta_1}]\right\}_A + \left\{\left(\frac{\varepsilon B}{\overline{G}}\right)\left(\frac{\overline{G}V}{\overline{\omega}_1}\right)\frac{e^{-\theta_1}}{x}\right\}_B \quad (20)$$

$$\overline{\mu}_1 = \left\{\left(\frac{\varepsilon B}{\overline{G}}\right)\left(\frac{V\overline{G}}{\overline{\omega}_1}\right)^2[(d\,r_0' + d^2) - e^{-\theta_1}(d\,r_c' + d^2)]\right\}_A + \left\{\left(\frac{\varepsilon B}{\overline{G}}\right)\left(\frac{V\overline{G}}{\overline{\omega}_1}\right)^2\left[e^{-\theta_1}\left(\frac{r_c'}{x} + \frac{1}{x^2}\right)\right]\right\}_B \quad (21)$$

where

$$d = \overline{\varepsilon}/[1 - x(1 - \overline{\varepsilon})] \quad (24)$$

$$\theta_1 = \left[\frac{1 - x(1 - \overline{\varepsilon})}{\overline{\varepsilon}}\right](r_c' - r_0') \quad (25)$$

The bracketed terms with subscript A are the contribution of fine crystals less the cutoff size r_C. The brackets B arise from crystals too large to be removed by the classifier. In Sherwin's case, with no fines drawoff, $r_c' \longrightarrow r_0' \longrightarrow 0$ and $\omega_3 \longrightarrow \omega_1$ $(x \longrightarrow 1)$. The $d \longrightarrow 1$, and $\exp(-\theta_1) \longrightarrow 1$. The A brackets vanish, and the B brackets reduce to Sherwin's expressions for small r_0'.

Figure 2 shows the steady state distribution for a particular case, both for no seeds in the feed stream and when seeds of one size r_s are added.

It was of interest to see what effect the second drawoff might have on the possible oscillatory behavior discussed by Sherwin. To establish limits of stable operation, the moment equations were linearized and examined by standard procedures for stability. Results for the case of completely clear mother liquor drawoff (no fines) are shown in Figure 3. It is evident that this drawoff has a stabilizing effect, but that instabilities are still possible in some conditions.

The design parameters are r_0' nucleus size x ratio of product drawoff to feed rate, ε fractional mother liquor volume in the magma, b nucleation sensitivity, and g growth sensitivity. In this work, we have assumed first-order growth kinetics:

$$G(c) = k_1 (c - c_s) \tag{26}$$

and

$$g = \frac{\bar{\varepsilon}_1 (1 - \bar{\varepsilon})(\bar{c}_1 - \bar{c})}{\bar{\varepsilon}(\bar{\varepsilon}_1 - \bar{\varepsilon})\bar{G}} \left.\frac{dG}{dc}\right|_{c=\bar{c}} = \frac{\bar{\varepsilon}_1 (1 - \bar{\varepsilon})(\bar{c}_1 - \bar{c})}{\bar{\varepsilon}(\bar{\varepsilon}_1 - \bar{\varepsilon})(\bar{c} - c_s)} \tag{27}$$

For no seeds in the feed, $\bar{\varepsilon}_1 = 1$. Generally, the steady state supersaturation $\bar{c} - c_s$ is much smaller than the difference of the inlet and outlet mother liquor concentrations. Thus, one might expect

$$10^2 < g < 10^3$$

The nucleation sensitivity b is

$$b = \frac{\bar{\varepsilon}_1 (1 - \bar{\varepsilon})(\bar{c}_1 - \bar{c})}{\bar{\varepsilon}(\bar{\varepsilon}_1 - \bar{\varepsilon})\bar{B}} \left.\frac{dB}{dc}\right|_{c=\bar{c}} \tag{28}$$

Taking the nucleation kinetics to follow the Volmer law

$$B(c) = k_2 \exp\{-k_3/[\ln(c/c_s)]^2\} \tag{29}$$

one finds

$$b = \frac{\bar{\varepsilon}_1 (1 - \bar{\varepsilon})(\bar{c}_1 - \bar{c})}{\bar{\varepsilon}(\bar{\varepsilon}_1 - \bar{\varepsilon})} \frac{2k_3}{\bar{c}[\ln(\bar{c}/c_s]^2} \tag{30}$$

When $\bar{c}/c_s$ is nearly unity (low exit supersaturation), Taylor series expansion of the logarithm gives

$$b = \frac{\bar{\varepsilon}_1 (1 - \bar{\varepsilon})}{\bar{\varepsilon}(\bar{\varepsilon}_1 - \bar{\varepsilon})} \frac{2k_3 c_s^3}{\bar{c}(\bar{c} - c_2)^3} \tag{31}$$

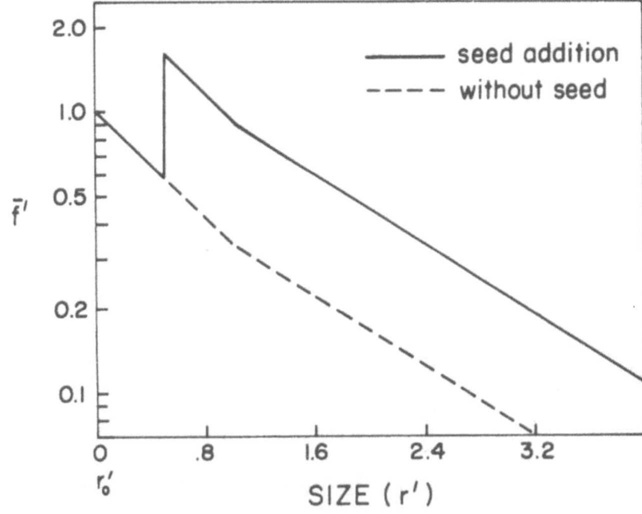

Fig. 2. Effect of seed addition on steady state size distribution.

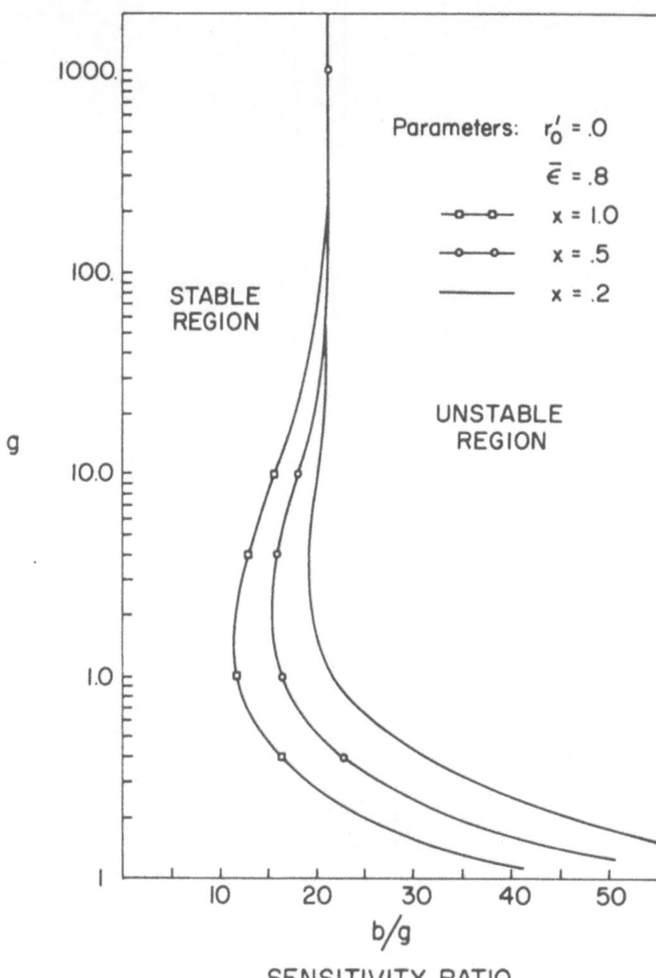

Fig. 3a. Stability criteria for noncyclic operation. No seed added.

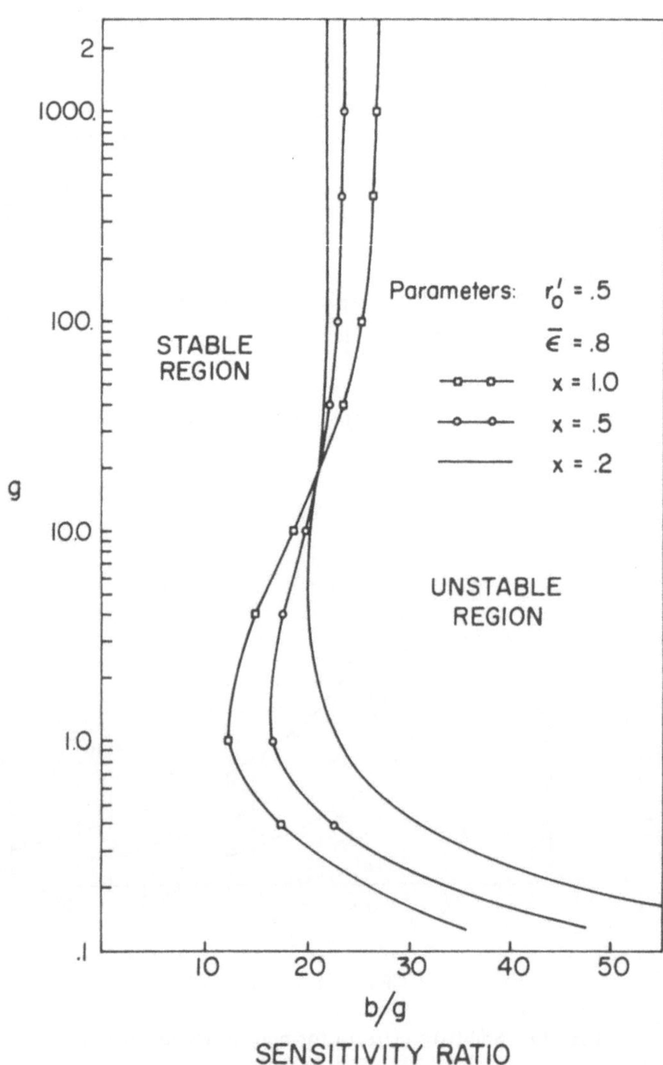

Fig. 3*b*. Stability criteria for noncyclic operation. Seed removal.

The sensitivity ratio is thus approximately

$$\frac{b}{g} = 2\, k_3 \Bigg/ \left(\frac{\bar{C} - C_s}{C_s}\right)^2 \tag{32}$$

It becomes large as the supersaturation becomes low in the crystallizer body.

For normally large growth rates, the stability limit is relatively insensitive to design parameters. Thus, for $g > 100$, operation is stable and steady for all $b/g < 21$ if nuclei are taken to have zero size. Other cases show that the limit shifts to larger b/g slightly if nuclei have a finite initial size, since the solids deposit rate is now increased in proportion to the nucleus area. But for expected nuclei sizes, this effect should be negligible. For extremely slow growth rates, or very close approach to saturation ($g < 1$), the stability limit becomes sensitive to the mother liquor drawoff rate, shifting to higher b/g values as the mother liquor drawoff increases. Thus, the second drawoff has a stabilizing effect.

In Figure 4, the effects of increased quantity of seed addition and increased seed size in the feed stream are both seen to increase the stability limit for $g > 100$.

To examine the nature of the instability more closely, numerical integration of the full set of nonlinear moment equations was carried out. In each case, it was found that the unstable condition was a limit cycle relaxation oscillation, attained more rapidly the further the operating conditions were from the stability boundary. These oscillations have a wave form and frequency which depend upon nucleation and growth kinetics as well as upon operating conditions, but no analytical expression has as yet been found for this dependence.

When fines are assumed to be present in the clarified mother liquor, the moment equations involve an assessment of the fines end of the total size distribution. They must thus be solved iteratively with Equation (5), even in the unsteady state case. To expedite this, we have assumed that the fines end is represented by a modified gamma distribution

$$f^A(r) = \frac{\lambda}{a}\, e^{-(\lambda r/a)} \left(\frac{\lambda r}{a}\right)^{\lambda} \frac{1}{(\lambda - 1)!} \{\mu_0 + k_3 \mu_3\} \tag{33}$$

in which the parameters λ, a, and k_3 are adjusted to reproduce μ_1, μ_2, and μ_3. The moments of $f(r)$ are expressible in terms of gamma functions. The moments of the fines end

$$\mu_n^A = \int_0^{r_c} r^n f^A(r)\, dr \tag{34}$$

are expressible in terms of incomplete gamma functions. Thus, with $f^A(r)$ determined so as to fit μ_0 through μ_3, the μ_n^A were calculated and the moment equations solved to give new estimates of μ_1, μ_2, and μ_3. Iteration at each time step continued until the partial moments of f^A agreed to sufficient accuracy with those formed from the moment equations of the previous iteration.

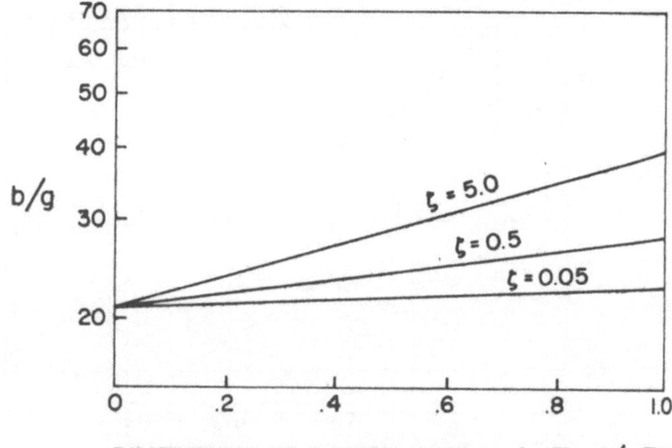

Fig. 4. Effect of seed addition on critical sensitivity ratio for noncyclic operation.

This result was compared in Figure 5*a* with a direct solution for the steady state case, which gives $f(r)$, and from this, the fines distribution $f^A(r)$. The partial moments of the fines were found to be from 5 to 10% low for a reasonable range of values of the parameters. However, since μ_n^A enters the moment equations only as a correction term, it was felt that this error was tolerable.

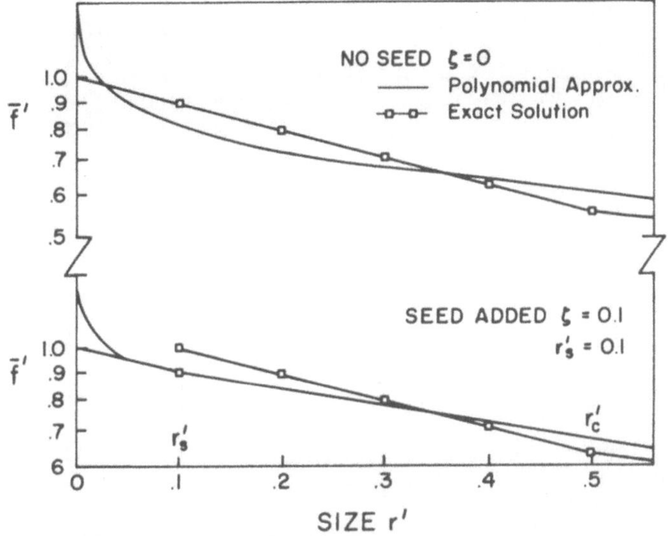

Fig. 5*a*. Approximate vs. exact steady state size distributions.

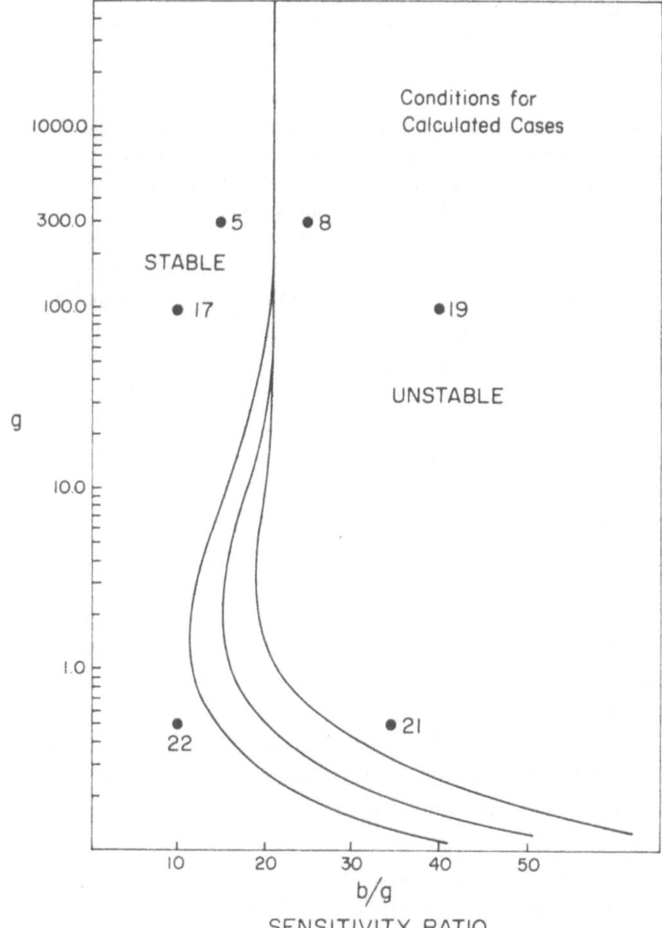

Fig. 5*b*. Operating conditions for numerical examples.

The analytical expression for μ_n^A by using $f^A(r)$ can be shown to be

$$\mu_n^A = \left(\frac{a}{\lambda}\right)^n \frac{\mu_0}{\Gamma(\lambda)} \left\{\left(\frac{\lambda r_c}{a}\right)^{\nu/2} \left(\frac{2}{\nu}\right)\left[1 - \left(\frac{\nu}{\nu+2}\right)\left(\frac{\lambda r_c}{a}\right)\frac{1}{1!} + \cdots + (-1)^k \left(\frac{\nu}{\nu+2k}\right)\left(\frac{\lambda r_c}{a}\right)^k \frac{1}{k!} + \cdots\right]\right\} + \left(\frac{a}{\lambda}\right)^n \frac{1}{\Gamma(\lambda)} \times \sum_{m=3}^{l}\left\{k_m \sum_{j=0}^{m}\left[\frac{(-1)^j m!\,\Gamma(m+\lambda)}{j!\,(m-j)!\,\Gamma(m+\lambda-j)} \times \left(\frac{\lambda r_c}{a}\right)^{v/2}\left(\frac{2}{v}\right) \times \left[1 - \left(\frac{v}{v+2}\right)\left(\frac{\lambda r_c}{a}\right)\frac{1}{1!} + \cdots + (-1)^k\left(\frac{v}{v+2k}\right)\left(\frac{\lambda r_c}{a}\right)^k \frac{1}{k!} + \cdots\right]\right]\right\} \quad (35)$$

where

$$\nu = 2(n + \lambda) \quad (36)$$

$$v = 2(m + n + \lambda - j) \quad (37)$$

Since k_m is a function of $\mu_0, \mu_1, \ldots, \mu_m$, the fines distribution is thus defined in terms of the first m moments of the full distribution.

Figures 6 to 8 give the response to an upset in a crystallizer initially operating at steady state induced by a sudden decrease of 20% in the clear mother liquor drawoff rate and a simultaneous increase of 20% in the magma drawoff rate. The ratio Z_n of the departure of the perturbed moment μ_n' from its initial value to the initial steady value $\bar{\mu}_n$ is shown as a function of elapsed time for a number of cases with initial stationary states chosen as shown in Figure 5*b*. Cases V and XXII are

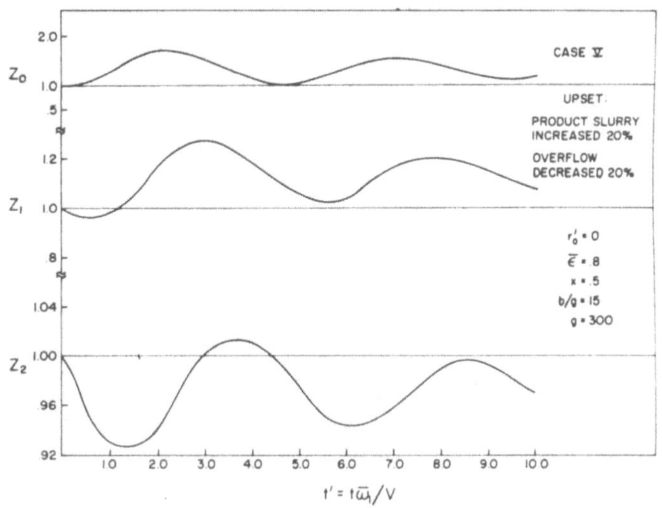

Fig. 6*a*. Ratio $Z_n = \mu_n'/\bar{\mu}_n$ of perturbed moment to steady state value after upset. Z_0 = No. density; Z_1 = mean size; Z_2 = mean area; Z_3 = magma density; Y = mother liquor concentration.

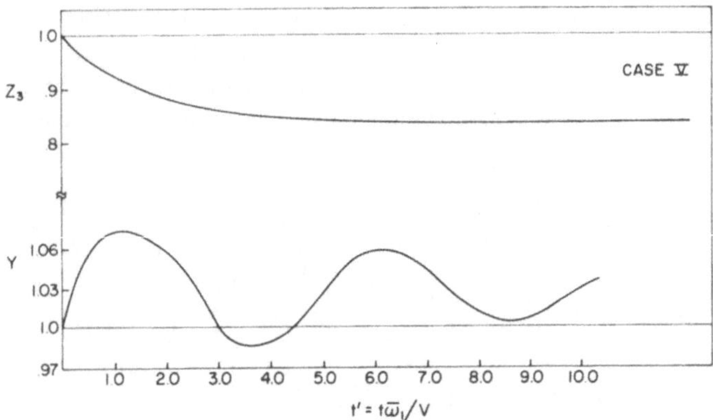

Fig. 6*b*. Ratio $Z_n = \mu_n'/\bar{\mu}_n$ of perturbed moment to steady state value after upset. Z_0 = No. density; Z_1 = mean size; Z_2 = mean area; Z_3 = magma density; Y = mother liquor concentration.

stable, and the perturbation in the moments consists of damped oscillations which die away fairly rapidly. The magma density Z_3 shows almost no oscillatory behavior at high g (case V), but the number and number mean size, Z_0 and Z_1 show oscillations for more than ten

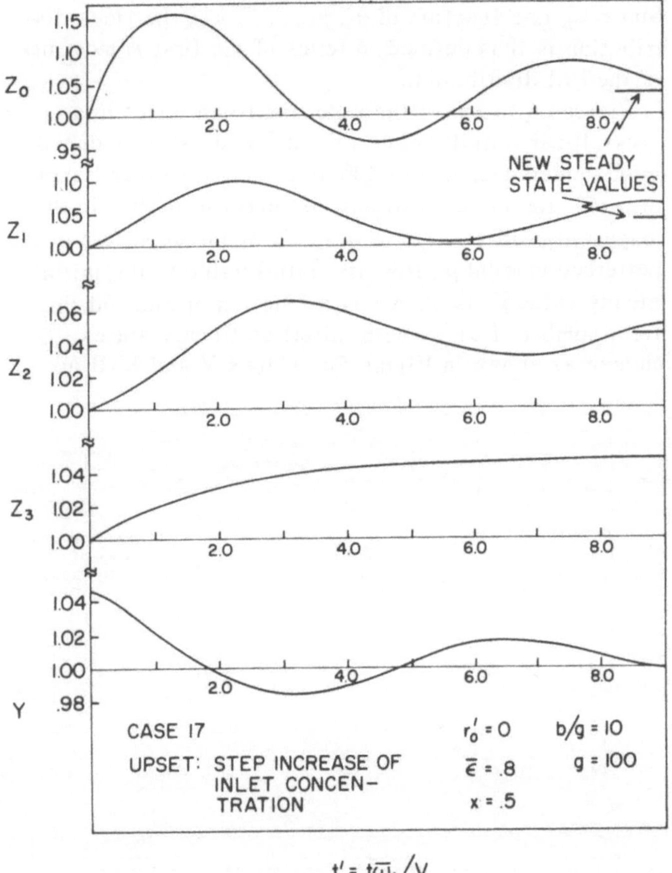

Fig. 7. Ratio $Z_n = \mu_n'/\bar{\mu}_n$ of perturbed moment to steady state value after upset. Z_0 = No. density; Z_1 = mean size; Z_2 = mean area; Z_3 = magma density; Y = mother liquor concentration.

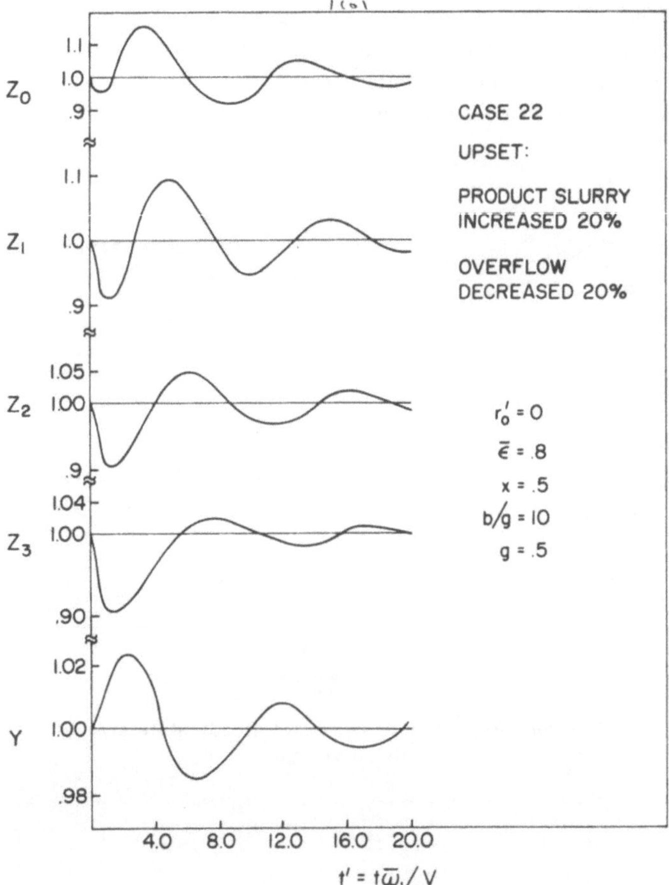

Fig. 8. Ratio $Z_n = \mu_n'/\bar{\mu}_n$ of perturbed moment to steady state value after upset. Z_0 = No. density; Z_1 = mean size; Z_2 = mean area; Z_3 = magma density; Y = mother liquor concentration.

residence times. When saturation is very closely approached (case XXII, small g), oscillations in magma density are apparent, but rapidly become small. In case XVII, the upset consists of a sudden increase in feed concentration, and again rapidly decaying oscillations are observed.

The unstable cases present the most interest. Figures 9 to 11 show the nonlinear oscillations that are computed. The initial oscillations now grow to a limiting amplitude and are quite unsymmetrical in time. Thus, Figure 11 (case XXI, g small) shows bursts of fines every ten residence times, which very shortly follow minima in the magma density and maxima in mother liquor concentration. At high growth rates (high exit supersaturation), Figure 9 (case VIII) shows less spectacular and more regular oscillation. The magma density, however, shows almost indetectable variation. Mother liquor concentration and mean size show large and rapid oscillations with a period of about four and one-half residence times. In case XIX, Figure 10, the upset is a feed concentration increase of 10%. Again, magma density oscillations are almost undetectable, but number density and mother liquor concentration oscillate

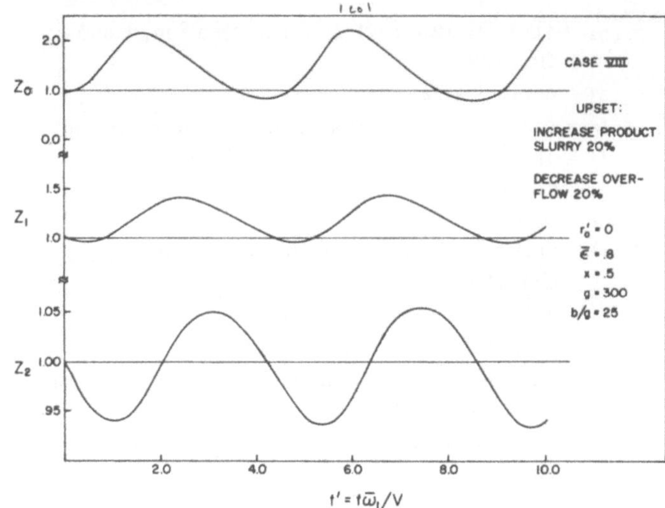

Fig. 9*a*. Ratio $Z_n = \mu_n'/\bar{\mu}_n$ of perturbed moment to steady state value after upset. Z_0 = No. density; Z_1 = mean size; Z_2 = mean area; Z_3 = magma density; Y = mother liquor concentration.

with large amplitudes with a period of about six and one-half residence times. In contrast to this, Figure 12, case XXVIII, shows the effect of taking off fines with the mother liquor. The classifier is taken to transmit only crystals less than $r_c' = 0.5$ (that is, smaller than could be grown in one-half a residence time). The instability persists, but the period is reduced to five residence times, and the amplitude is also much smaller. When the fines size is doubled in the drawoff, the period is reduced further to two and one-half residence times with little further effect on the amplitude.

When much smaller fines are retained, case XXXV; the period is about five and one-half residence times. In all cases, with fines drawoff the fines seem to burst suddenly into appearance within one residence time. This is especially evident in $Z_0{}^A$, the normalized number of particles in the drawoff, which oscillates from zero to

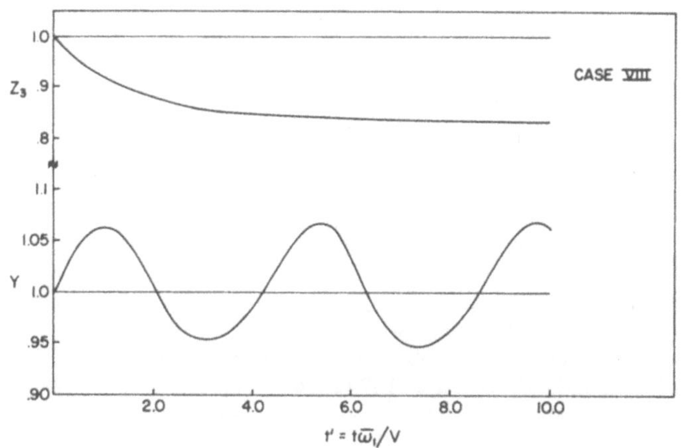

Fig. 9b. Ratio $Z_n = \mu_n'/\bar{\mu}_n$ of perturbed moment to steady state value after upset. Z_0 = No. density; Z_1 = mean size; Z_2 = mean area; Z_3 = magma density; Y = mother liquor concentration.

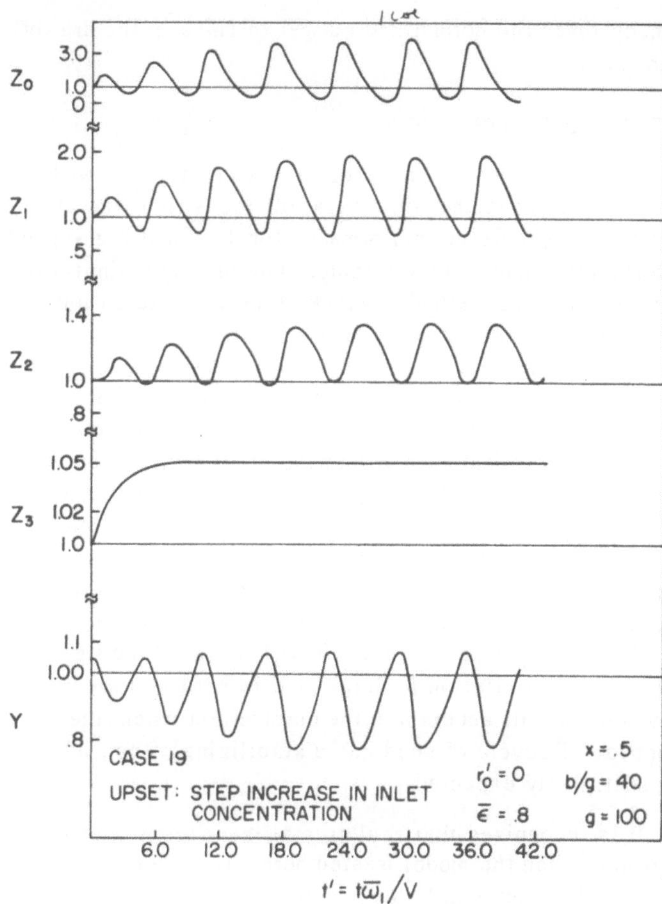

Fig. 10. Ratio $Z_n = \mu_n'/\bar{\mu}_n$ of perturbed moment to steady state value after upset. Z_0 = No. density; Z_1 = mean size; Z_2 = mean area; Z_3 = magma density; Y = mother liquor concentration.

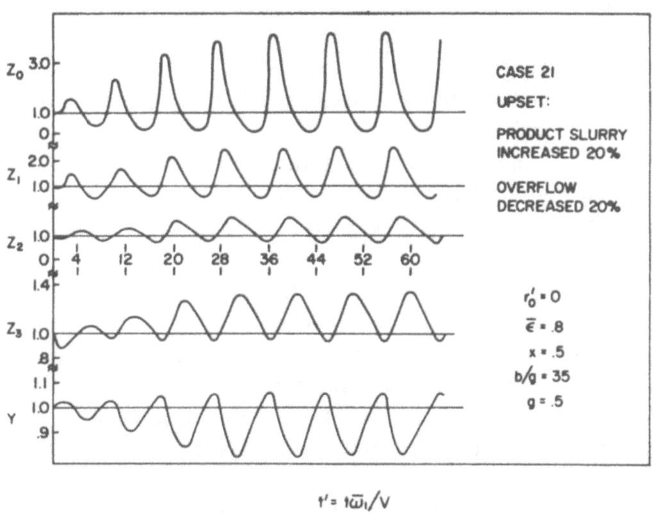

Fig. 11. Ratio $Z_n = \mu_n'/\bar{\mu}_n$ of perturbed moment to steady state value after upset. Z_0 = No. density; Z_1 = mean size; Z_2 = mean area; Z_3 = magma density; Y = mother liquor concentration.

many times the normalized number of fines in the drawoff initially.

CONCLUSIONS

The simple model of a stirred-tank crystallizer with fines or clear mother liquor drawoff leads to a tractable method of predicting performance for design if high-speed computer capacity is available. The principal limitation in use of these methods is lack of data on nucleation and growth rates to assess the stability criteria and the mean performance. However, the qualitative sensitivity of the design to assumed parameter values can be explored and safe values chosen. The most significant feature of oscillating performance is its long period, easily a number of hours or days. This has undoubtedly led to operating problems of obscure origin in the past. It is also significant that crystal production rate shows negligibly small oscillations under conditions where there are large swings in crystal size and numbers.

Addition of a second drawoff will increase the frequency of oscillation as larger and larger fines are removed but will accentuate the burst effect when fines do appear. Recycle of seed has a stabilizing effect, as qualitatively expected.

It is recognized that real crystallizers are much more complex than the model treated here. However, to the extent that they can be considered as interlinked zones of uniform conditions, with specifiable transfer rates between the zones, the methods used here can be extended at only the price of increased computation time.

NOTATION

a = parameter of gamma distribution, Equation (33)
$B(c)$ = number of nuclei formed per unit clear liquor volume per unit time
b = nucleation sensitivity factor, Equation (28)
c = mother liquor concentration, mass per unit volume of clear liquor
c_s = concentration of saturated mother liquor
d = group defined in Equation (24)
$f(r, x; t)$ = number fraction of crystals per unit volume per unit size
f_w = weight fraction of crystals per unit volume per unit size, Equation (11)
$G(c)$ = crystal growth rate, volume deposited per unit surface area per unit time
g = growth sensitivity factor, Equation (27)
$h(x, r; t)$ = net rate of production of crystals of size r to $r + dr$
k_1 = specific growth rate, Equation (26)
k_2, k_3 = specific rate factors for nucleation, Equation (29)
r = crystal size
r_c = size of largest crystal in discharge stream
r_j' = dimensionless crystal size, $\omega_1 g/VG$, in stream j
t = time
$U(x)$ = unit step function, defined following Equation (19)
V = volume of crystallizer body
v = term defined in Equation (37)
$x = \bar{\omega}_3/\bar{\omega}_1$, ratio of product slurry volume flow to feed volume flow
Z_n = ratio of perturbed moment μ_n' to its initial value

GREEK LETTERS

$\Gamma(\lambda)$ = gamma function
ξ_j = j^{th} moment of particle size distribution in feed stream
θ_1 = term defined in Equation (25)
λ = parameter of gamma distribution, Equation (33)
μ_j = j^{th} moment of size distribution, Equations 16) to 18)
μ_j' = perturbed moment
ν = term defined in Equation (36)
ν_j = j^{th} moment of particle size distribution in mother liquor and fines drawoff
φ = density of crystal solid
σ = crystal surface per unit volume of crystallizer
Φ = shape-density factor, Equation (10)
ω_j = volume discharge rate of stream j

SUBSCRIPTS

0 = crystal nuclei
1 = feed stream
2 = mother liquor and fines drawoff
3 = main product stream

SUPERSCRIPTS

A = fines distribution
—(bar) = average values or steady state values
′(prime) = dimensionless variables or perturbed moments

LITERATURE CITED

1. Hulburt, H. M., and T. Akiyama, *Am. Chem. Soc. Chem. Eng. Symposium*, Boston, Mass. (1967).
2. Hulburt, H. M., and S. Katz, *Chem. Eng. Sci.*, **19**, 555 (1964).
3. Murray, D. C., and M. A. Larson, *AIChE J.*, **11**, 728 (1965).
4. Randolph, A. D., *ibid.*, 424.
5. ———, and M. A. Larson, *ibid.*, **8**, 639 (1962).
6. Saeman, W. C., *ibid.*, **2**, 107 (1956).
7. Sherwin, M. B., Reuel Shinnar, and Stanley Katz, *ibid.*, **13**, 1141 (1967).

DYNAMIC BEHAVIOR OF THE WELL-MIXED ISOTHERMAL CRYSTALLIZER

M. B. Sherwin
Reuel Shinnar
and
Stanley Katz

Under some conditions, continuous crystallization exhibits cyclic changes of the particle size even with constant input conditions. A linearized stability analysis has been performed to determine under what conditions this behavior can be expected. Numerical solutions of the actual nonlinear system equations were carried out to follow the cyclic behavior and to compute the cycle time and the amplitude of the fluctuations. The effect of seeding on the stability limits and product distribution was also evaluated.

Industrial crystallization operations are still something of an art, and often depend to a considerable degree on the skill of the operator. This paper attempts to aid in the understanding of the dynamic behavior and stability of such systems.

The special feature which distinguishes continuous crystallization processes from other continuous reactors is the simultaneous occurrence of nucleation and growth. Both these steps depend on supersaturation as a driving force, with the rate dependence of nucleation on supersaturation being highly nonlinear. In a well-mixed crystallizer these steps occur simultaneously and homogeneously throughout the vessel.

It has been observed that continuous crystallization processes are under some conditions inherently unstable and of a cyclic nature (*10*). Randolph and Larson (*12*) speculated that the instability and long-term transients in the size distribution observed in an ammonium sulfate crystallizer were due to changes in operation over a three-shift period rather than inherent to the system. Saeman (*10*) found that it was impossible to operate a completely classified ammonium nitrate crystallizer continuously, but was forced to discharge periodically product in a cyclic manner. Changing the operation to a well-mixed crystallizer alleviated, but did not completely remove, the operating problems. Much of this unsteady behavior occurs despite the fact that heat and material inputs and other operating conditions are held constant. Similar observations have been made in some continuous polymerization and fermentation processes which also involve nucleation and growth (*6, 18*).

Theoretical papers on industrial crystallization are very few in number and deal mainly with steady state behavior (*2, 15*). Experimental and theoretical studies on nucleation or growth usually focus on just one of these kinetic steps, thereby avoiding the problem of interaction. This study of the dynamics of the well-stirred crystallizer considers these interactions and tries to analyze what effect variation in operation will have on them. An explanation is proposed for the observed difficulty of attaining continuous stable operation in some cases and means of correcting this behavior become apparent from the text.

Although considerable effort has been spent on the study of the dynamical behavior of both homogeneous and heterogeneous reaction systems (*1, 4, 21*), very few studies deal with the effect of nucleation. Part of this is due to the extreme difficulty in the mathematical treatment of such systems in all but the most simple cases.

Larson (*12*) has made a theoretical linearized stability analysis of a completely mixed crystallizer, and found that under most practical conditions it should be stable. But in an experimental step response study of one typical crystallization, he also found strong tendency toward long-term cyclic transients.

Larson's analysis contains some very strong simplifications, particularly in the neglect of a metastable region of low nucleation rate. It is just this strong nonlinear dependence of nucleation rate on supersaturation that causes the observed cyclic changes in particle size. Our linearized stability analysis allows for such effects, and hence shows both regions of stable and unstable operation.

A linearized stability analysis will only indicate if the system is stable, and will not give any indication as to the magnitude of the fluctuations. Fortunately, for a mixed vessel it is in many cases possible to follow the behavior of a disturbance in the nonlinear region and to compute the amplitude of the fluctuations.

The approach used in this work was presented in a recent paper by Katz and Hulburt (*8*) which allowed a reasonably realistic, theoretical treatment for quite complex situations.

DERIVATION OF THE MODEL

A diagram of the type of operation discussed is shown in Figure 1. A feed material containing solute at concentra-

M. B. Sherwin is with Chem Systems, Inc., New York, New York.

Reprinted from the November, 1967, issue of the AIChE Journal, Volume 13, No. 6 (pages 1141–1154).

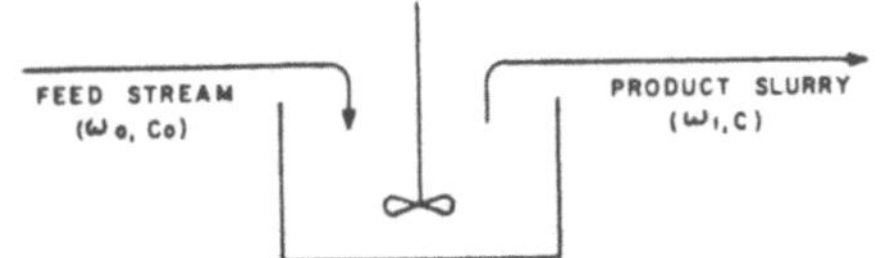

Fig. 1. Mixed crystallizer with mixed-product removal.

tion c_0 (moles per unit volume) is fed to a crystallizer of working volume V at a volumetric rate ω_0 (volume per unit time). The crystallizer is well mixed and isothermal so that instant cooling of the feed results in a supersaturation driving force for subsequent nucleation and growth. A product slurry representative of the contents of crystallizer body is withdrawn at volumetric rate ω_1.

In order to describe the behavior of the isothermal mixed suspension mixed product removal crystallizer, balances must be written for the particles and the crystallizing material. In addition to these general balances, appropriate kinetic relations must be added for the description to be complete.

In writing these balances the following assumptions were made:

The crystal particles are assumed to be regular solids which maintain their shape upon growth and can therefore be characterized geometrically by a linear dimension r.

For solids which grow in three dimensions, the volume of any crystal is given by Kr^3, where K is the appropriate geometrical shape factor. (This paper will deal with the case of a three-dimensional crystal since it is the most common, although the assumption of plate or needle shaped crystals, which grow in two- and one-dimensions, respectively, could just as easily be substituted.)

The crystal density ρ is taken to be constant.

The particle size distribution is the number density f at size r, such that $f(r, t)\,dr$ = the number of crystals per unit volume of crystallizer having radii in the range r, $r + dr$ at time t.

The fractional volume ε occupied by the solution is a function of time and can be defined in terms of f

$$1 - \varepsilon(t) = \int_0^\infty Kr^3 f(r, t)\,dr \tag{1}$$

Similarly, when a seeded feed is used, $\psi(r, t)\,dr$ = the number of crystals per unit volume of feed having radii in the range r, $r + dr$ at time t and the fractional volume ε occupied by the feed solution is

$$1 - \varepsilon_0(t) = \int_0^\infty Kr^3 \psi(r, t)\,dr \tag{2}$$

The growth rate of an existing crystal dr/dt is assumed to be described by the product of a function of crystal environment $G(c)$ and a function of crystal size $\phi(r)$ such that.

$$\frac{dr}{dt} = G(c)\,\phi(r) \tag{3}$$

where c is the solute concentration per volume of solution.

The nucleation rate for new crystals is mainly a function of solute concentration in the crystallizer according to the classical theories of Volmer (*19*) and Mier (*20*). It is assumed that the size distribution of new crystals appears as a Dirac delta function, or that all new crystals are formed at a nominal size r_0.

$B(c)$ = number of crystals born per volume of solution per unit time at r_0. ω_0 and ω are the volumetric feed and withdrawal rates to and from the crystallizer working volume V.

Conservation Equations

Now the appropriate balances over the crystallizer can be written down. (A detailed derivation of the particle balance can be found in reference *8*.)

Particle balance:

$$\frac{\partial[Vf]}{\partial t} + \frac{\partial[G\phi Vf]}{\partial r} = \varepsilon VB\delta(r - r_0) - \omega_1 f + \omega_0 \psi \tag{4}$$

These terms represent, in order, the accumulation of crystals at size r, the net flux of crystals away from size r due to growth, the input of particles at size r_0 due to nucleation, the withdrawal of particles of size r due to product removal, and the input of particles of size r due to solids in the feed.

Solute and crystal balance:

$$\frac{d}{dt}[V\{\varepsilon c + (1 - \varepsilon)\,\rho\}] = \omega_0[\varepsilon_0 c_0 + (1 - \varepsilon_0)\rho] - \omega_1[\varepsilon c + (1 - \varepsilon)\rho] \tag{5}$$

These terms represent, in order, the accumulation of solute and crystal, the input of solute and crystal by the feed stream, and the removal of solute and crystal due to product withdrawal.

When Equation (4) is multiplied by Kr^3 and integrated over r, the following relation is obtained:

$$\frac{d}{dt}[(1 - \varepsilon)V] = GV\sigma + \varepsilon VBKr_0^3 + \omega_0(1 - \varepsilon_0) - \omega_1(1 - \varepsilon) \tag{6}$$

where

$$\sigma = 3K\int_0^\infty r^2 f\phi\,dr \tag{7}$$

If a balance were written for the solvent and this relation combined with (5) and (6), one could obtain an expression for the variation of the working volume V with time. The resulting expression is

$$\frac{dV}{dt} = (\omega_0 - \omega_1) + (1 - \bar{v}\rho)(\sigma VG + \varepsilon VBKr_0^3) \tag{8}$$

where $\bar{v}$ is the partial molar volume of the solute. Available data indicate that taking $\bar{v} = 1/\rho$ is quite a reasonable assumption, so that if we specify that $\omega_0 = \omega_1 = \omega$, the working volume remains constant.

It is desirable to have a relation expressing the variation of solute concentration c with time, since both G and B are expressed in terms of this variable. Equations (5) and (6) can be used along with the assumption of constant volume to yield the desired expression:

$$\varepsilon\frac{dc}{dt} = \frac{\omega}{V}(c_0 - c)\,\varepsilon_0 - (\rho - c)\,(\sigma G + \varepsilon BKr_0^3) \tag{9}$$

The working set of equations is now

$$\frac{\partial f}{\partial t} + \frac{\partial}{\partial r}[G\phi f] = \varepsilon B\delta(r - r_0) + \frac{\psi}{\theta} - \frac{f}{\theta} \tag{10}$$

$$\varepsilon\frac{dc}{dt} = \frac{(c_0 - c)\,\varepsilon_0}{\theta} - (\rho - c)(\sigma G + \varepsilon BKr_0^3) \tag{11}$$

where

$$\theta = \frac{V}{\omega}$$

$$\sigma = 3K\int_0^\infty r^2 f\phi\,dr$$

Moment Equations

The above set of equations is difficult to solve, even numerically, since it contains a partial differential equation for $f(r, t)$. If it is not necessary to determine how the distribution f varies with time, but a knowledge of the variation of its moments with time is satisfactory, the set can be transformed to a group of ordinary differential equations. The nth moment of f will be called μ_n while the expressions in pointed brackets are averages over f in the same sense; in general

$$\langle \eta \rangle = \int_0^\infty \eta f\, dr \tag{12}$$

$$\mu_n(t) = \int_0^\infty r^n f(r, t)\, dr \tag{13}$$

The physical significance of the moments of f is shown below:

$$\mu_0 = \int_0^\infty f\, dr = \text{total number of crystals per unit volume of crystallizer}$$

$$\mu_1 = \int_0^\infty r f\, dr = \text{total radius of crystals per unit volume of crystallizer}$$

$$3K\mu_2 = 3K \int_0^\infty r^2 f\, dr = \text{total surface of crystals per unit volume of crystallizer}$$

$$K\mu_3 = K \int_0^\infty r^3 f\, dr = (1 - \varepsilon) = \text{volume fraction of solids}$$

In the same manner, the moments of the seed crystal distribution can be written as

$$\eta_n(t) = \int_0^\infty r^n \psi(r, t)\, dr \tag{14}$$

and these moments retain the same physical significance as the corresponding moments related to f per unit volume of feed.

The weight distribution of the crystals $W(r)$ can be written as

$$W(r) = \rho K r^3 f(r) \tag{15}$$

so that the mean particle size with respect to the weight distribution is

$$\text{Mean particle size} = r_{wm} = \frac{\int_0^\infty r W(r)\, dr}{\int_0^\infty W(r)\, dr} = \frac{\rho K \int_0^\infty r^4 f(r)\, dr}{\rho K \int_0^\infty r^3 f(r)\, dr} = \frac{\mu_4}{\mu_3} \tag{16}$$

The variance of the weight distribution is

$$\text{Variance} = \frac{\int_0^\infty (r - r_{wm})^2 W(r)\, dr}{\int_0^\infty W(r)\, dr} =$$

$$\frac{\int_0^\infty r^2 W(r)\, dr - 2r_{wm} \int_0^\infty W(r)\, dr + r_{wm}^2 \int_0^\infty W(r)\, dr}{\int_0^\infty W(r)\, dr} \tag{17}$$

which, in terms of the number distribution is

$$\text{Variance} = \frac{\mu_5}{\mu_3} - \left(\frac{\mu_4}{\mu_3}\right) \tag{18}$$

The coefficient of variation γ of the weight distribution is a measure of its uniformity.

$$\gamma^2 = \frac{\text{variance}}{(\text{mean size})^2} = \frac{\frac{\mu_5}{\mu_3} - \left(\frac{\mu_4}{\mu_3}\right)^2}{\left(\frac{\mu_4}{\mu_3}\right)^2} \tag{19}$$

Clearly all these moments in Equations (16), (17), and (19) can be determined by a microscopic analysis of a sieve analysis of the crystallizer product. It is also clear that a knowledge of these moments is enough to characterize the performance of a crystallizer.

By multiplying Equation (10) through by r^n and integrating, one can generate the moment equations:

$$\frac{d\mu_n}{dt} = nG(c) \langle r^{n-1} \phi(r) \rangle + \varepsilon B(c) r_0{}^n + \frac{\eta_n}{\theta} - \frac{\mu_n}{\theta}$$

$$n = 0, 1, 2, 3 \dots \tag{20}$$

where

$$\langle r^{n-1} \phi(r) \rangle = \int_0^\infty r^{n-1} \phi(r) f\, dr$$

Now the series of Equations (20) plus Equation (11) is the set of equations describing the system. These, however, do not necessarily give a self-contained set of equations because of the dependence of particle growth on r, as manifested in the terms $\langle r^{n-1} \phi \rangle$. If dr/dt is independent of r, then $\phi = 1$ and $\langle r^{n-1} \phi \rangle$ is simply μ_{n-1} and the equations close satisfactorily at $n = 3$. This is so even if dr/dt depends linearly on r, but more awkward terms of dependence (r^2, r^3, etc.) can only be handled by making successive approximations to $f(r, t)$ in the term of Laguerre series as discussed in reference *8*.

Fortunately a number of investigators (*9, 14, 17*) have found that in many cases the growth rate does not depend on r to any significant amount, so that in the following sections it is assumed that $\phi = 1$.

Later in the paper growth models of the form

$$\frac{dr}{dt} = G(c)[1 + ar] \tag{21}$$

which include a size-dependent term will be studied to illustrate the effect of this dependence.

Remembering that $1 - \varepsilon = \int_0^\infty K r^3 f\, dr$, we get the following closed set of equations:

$$\frac{d\mu_0}{dt} = (1 - K\mu_3) B(c) - \frac{\mu_0}{\theta} + \frac{\eta_0}{\theta}$$

$$\frac{d\mu_1}{dt} = G(c)\mu_0 + (1 - K\mu_3) B(c) r_0 - \frac{\mu_1}{\theta} + \frac{\eta_1}{\theta}$$

$$\frac{d\mu_2}{dt} = 2G(c)\mu_1 + (1 - K\mu_3) B(c) r_0^2 - \frac{\mu_2}{\theta} + \frac{\eta_2}{\theta} \qquad (22)$$

$$\frac{d\mu_3}{dt} = 3G(c)\mu_2 + (1 - K\mu_3) B(c) r_0^3 - \frac{\mu_3}{\theta} + \frac{\eta_3}{\theta}$$

$$(1 - K\mu_3)\frac{dc}{dt} = \frac{(c_0 - c)}{\theta}\varepsilon_0 - (\rho - c)(3G(c)K\mu_2 + (1 - K\mu_3)B(c)Kr_0^3)$$

In order to solve this set of equations to trace the history of $c(t)$ and the moments of $f(r, t)$, the system (22) must be supplemented by the initial conditions and the feed rate and composition as a function of time. Also the dependence of $B(c)$ and $G(c)$ on concentration $c(t)$ must be specified.

Steady State Equations

In order to get the steady state moments, the particle distribution $\overline{f}(r)$ must first be obtained. This can then be used to generate the moment relations.

The steady state particle balance can be written down from (10), remembering that we are now assuming $\phi = 1.0$:

$$G(c)\frac{df}{dr} = \varepsilon\beta(c)\delta(r - r_0) - \frac{f(r)}{\theta} + \frac{\psi(r)}{\theta} \qquad (23)$$

where the superscript bars indicate steady state values.

This can be solved, giving

$$\overline{f} = \frac{\varepsilon B}{G} e^{-\frac{1}{\theta G}(r - r_0)} + \frac{1}{\theta G}\int_0^r \psi(x) e^{-\frac{1}{\theta G}(r - x)} dx \qquad (24)$$

The moments can now be obtained by straightforward calculation from (24):

$$\overline{\mu}_0 = \frac{\varepsilon B}{G}(\theta G) + \eta_0$$

$$\overline{\mu}_1 = \frac{\varepsilon B}{G}(\theta G)^2\left[\left(\frac{r_0}{\theta G}\right) + 1\right] + \theta G\eta_0 + \eta_1$$

$$\overline{\mu}_2 = \frac{\varepsilon B}{G}(\theta G)^3\left[\left(\frac{r_0}{\theta G}\right)^2 + 2\left(\frac{r_0}{\theta G}\right) + 2\right] + 2(\theta G)^2\eta_0 + 2\theta G\eta_1 + \eta_2 \qquad (25)$$

$$\overline{\mu}_3 = \frac{\varepsilon B}{G}(\theta G)^4\left[\left(\frac{r_0}{\theta G}\right)^3 + 3\left(\frac{r_0}{\theta G}\right)^2 + 6\left(\frac{r_0}{\theta G}\right) + 6\right] + 6(\theta G)^3\eta_0 + 6(\theta G)^2\eta_1 + 3(\theta G)\eta_2 + \eta_3 \quad \text{etc.}$$

An additional relation which will be of use later is derived from the last two relations in set (22).

$$\overline{\varepsilon} = 1 - K\overline{\mu}_3 = \left(\frac{\rho - \overline{c}_0}{\rho - \overline{c}}\right)\overline{\varepsilon}_0 \qquad (26)$$

LINEARIZATION OF THE MODEL

In order to determine under what conditions instability in the system will occur, the set of equations describing the system (22) must be linearized about the steady state. Then the conditions under which a small displacement from the steady state will grow may be determined via the conventional use of the Routh-Horwitz criterion. Primed quantities indicate the departure from the steady state values (superscript bar). The feed rate and concentration can vary in time in the form

$$\omega(t) = \overline{\omega} + \omega'(t)$$
$$c_0(t) = \overline{c_0} + c_0'(t) \qquad (27)$$

The initial conditions are

$$c(0) = \overline{c}$$
$$\mu_n(0) = \overline{\mu}_n \quad n = 0, 1, 2, \ldots \qquad (28)$$

The performance variables are represented by

$$\mu_n(t) = \overline{\mu}_n + \mu_n' \quad n = 0, 1, 2, \ldots$$
$$c(t) = \overline{c} + c'(t) \qquad (29)$$

The kinetic terms for small displacements can be written as

$$B(c) = \overline{B} + \frac{dB}{dc}(c)\cdot c'$$
$$G(c) = \overline{G} + \frac{dG}{dc}(c)\cdot c' \qquad (30)$$

In addition, the following dimensionless variables are defined:

$$g = \frac{\overline{\varepsilon}_0(1 - \overline{\varepsilon})}{\overline{\varepsilon}(\overline{\varepsilon}_0 - \overline{\varepsilon})}\frac{\overline{c}_0 - \overline{c}}{\overline{G}}\frac{dG}{dc}(\overline{c}) \qquad (31)$$

$$b = \frac{\overline{\varepsilon}_0(1 - \overline{\varepsilon})}{\overline{\varepsilon}(\overline{\varepsilon}_0 - \overline{\varepsilon})}\frac{\overline{c}_0 - \overline{c}}{\overline{B}}\frac{dB}{dc}(\overline{c}) \qquad (32)$$

where

$$\varepsilon = 1 - K\mu_3$$
$$\varepsilon_0 = 1 - K\eta_3$$
$$Z_n' = \frac{\mu_n'}{\overline{\mu}_n}$$
$$Y' = \frac{\overline{\varepsilon}(\overline{\varepsilon}_0 - \overline{\varepsilon})}{\overline{\varepsilon}_0(1 - \overline{\varepsilon})}\frac{c'}{\overline{c}_0 - \overline{c}} \qquad (33)$$
$$\tau = t\overline{\omega}/V = t/\theta$$
$$p(\tau) = \frac{\omega'(\tau)}{\overline{\omega}}$$
$$q(\tau) = \frac{(\overline{\varepsilon}_0 - \overline{\varepsilon})}{(1 - \overline{\varepsilon})}\frac{c_0'(\tau)}{\overline{c}_0 - \overline{c}}$$

Now substitution of Equations (27) to (33) into (22) and utilization of the steady state relations result in the following dimensionless and linearized set:

$$\frac{dZ_n'}{d\tau} - \left[\overline{\left(n\frac{\mu_{n-1}}{\mu_n}\theta G\right)}g + \overline{\left(1 - \frac{\eta_n}{\mu_n}\right)}b - \overline{\left(n\frac{\mu_{n-1}}{\mu_n}\theta G\right)}b\right]Y - \left[\overline{n\frac{\mu_{n-1}}{\mu_n}\theta G}\right]Z_{n-1} + \left[\overline{Kr_0 nB\theta\frac{\mu_3}{\mu_n}}\right]Z_3 + Z_n = -p(\tau)\left[\overline{1 - \frac{\eta_n}{\mu_n}}\right] \quad n = 0, 1, 2, \ldots.$$

with $Z'_n(0) = 0$

$$\frac{dY'}{d\tau} + \left[\left(\overline{3\frac{\mu_2}{\mu_3}\theta G}\right)g + \left(\overline{1 - \frac{\eta_3}{\mu_3}}\right)b - \left(\overline{3\frac{\mu_2}{\mu_3}\theta G}\right)b\right]Y' + Y' +$$

$$\left[\overline{3\frac{\mu_2}{\mu_3}\theta G}\right]Z'_2 - \overline{[Kr_0{}^3B\theta]}\,Z'_3 = q(\tau) + \left[\overline{1 - \frac{\eta_3}{\mu_3}}\right]p(\tau) \quad (34)$$

with $Y'(0) = 0$

There are three sets of dimensionless groups containing steady state variables which arise in (34):

$$L_n = n\frac{\overline{\mu_{n-1}}}{\overline{\mu}_n}\overline{\theta G} \quad n = 0, 1, 2, \ldots \quad (35)$$

$$R_n = Kr_0{}^3\overline{B}\overline{\theta}\frac{\overline{\mu_3}}{\mu_n} \quad (36)$$

$$S_n = \frac{\overline{\eta}_n}{\overline{\mu}_n} \quad (37)$$

besides the dimensionless kinetic parameters b and g which have been defined in Equations (31) and (32).

These sets of dimensionless groups L_n, R_n, and S_n depend on a finite number of steady state parameters contained in Equation (25). The identification and dependence of the model on these parameters will be discussed in succeeding sections.

KINETIC MODELS

Some well-known growth and nucleation models are now briefly introduced and substituted into Equations (31) and (32) in order to relate these important dimensionless kinetic parameters to the variables of the system.

It should be mentioned here that the later sections will show that the ratio of these two groups, b/g, is of primary importance in determining the stability and dynamic behavior of the model.

Growth Model

Many investigators have found the relation between G and c to be satisfied by

$$G(c) = K_1(c - c_s) \quad (38)$$

where K_1 is a constant for isothermal operation and c_s is the solubility concentration of the solute. Substitution of this model into Equation (31) yields

$$g = \frac{\overline{\varepsilon}_0(1 - \overline{\varepsilon})}{\overline{\varepsilon}(\overline{\varepsilon}_0 - \overline{\varepsilon})}\frac{(\overline{c}_0 - \overline{c})}{(\overline{c} - c_s)} \quad (39)$$

Generally, the supersaturation $(c - c_s)$ of crystallizing systems is quite small when compared with the difference between feed and outlet concentrations, so that normally the value of g should be of the order of magnitude 10^2 to 10^3.

Nucleation Models

Volmer's model (*15*) is based on thermodynamic considerations and gives rise to an Arrhenius type of function:

$$B = K_2 e^{\frac{K_3}{(\ln c/c_s)^2}} \quad (40)$$

Substitution of this model into Equation (32) yields

$$b = \frac{\overline{\varepsilon}_0(1 - \overline{\varepsilon})}{\overline{\varepsilon}(\overline{\varepsilon}_0 - \overline{\varepsilon})}\frac{(\overline{c}_0 - \overline{c})\,2K_3}{\overline{c}[\ln \overline{c}/c_s]^2} \approx \frac{\overline{\varepsilon}_0(1 - \overline{\varepsilon})}{\overline{\varepsilon}(\overline{\varepsilon}_0 - \overline{\varepsilon})} \times$$

$$\frac{\overline{c}_0 - \overline{c}}{\overline{\varepsilon}}\frac{2K_3}{\overline{c}(\overline{c}/c_s - 1)^3} \quad (41)$$

the latter approximation being made since $(c/c_s - 1)$ is very small for crystallization systems.

Then using the kinetic models given by Equations (38) and (40), we can approximate the term b/g by

$$\frac{b}{g} = \frac{2K_3(\overline{c} - c_s)}{(\overline{c}/c_s - 1)^3\overline{c}} = \frac{2K_3 c_s}{(\overline{c}/c_s - 1)^2\overline{c}} \approx \frac{2K_3}{(\overline{c}/c_s - 1)^2} \quad (42)$$

The term b/g represents the sensitivity ratio of the nucleation to growth functions at the steady state conditions and it can be seen that, as the solute concentration approaches its solubility limit, this term rapidly increases in value.

Another nucleation model which is a simplified representation of Equation (40) is Mier's metastable model (*16*) which can be written as

$$\begin{aligned} B &= K_4(c - c_m)^m \quad c > c_m \\ B &= 0 \quad\quad\quad\quad\quad\;\; c \leq c_m \end{aligned} \quad (43)$$

where k_4 is a constant, c_m is the metastable concentration, above which nucleation occurs, and m is some power. Combining this with Equation (38) we get

$$\frac{b}{g} = m\frac{\overline{c} - c_s}{\overline{c} - c_m} \quad c > c_m \quad (44)$$

Here it is evident that as the solute concentration approaches the metastable limit c_m from above, the sensitivity ratio b/g will increase in value.

It should be pointed out that nucleation in a mixed crystallizer is a highly complex phenomenon involving both homogenous and heterogenous nucleation, the latter being the preferential formation of nuclei near a surface of the same crystals. As Equation (40) was derived for homogenous nucleation one could therefore question its application to nucleation in a crystal magma.

In the strict theoretical sense, Vollmer's equation with its theoretically derived constants is not applicable. In the same way no data obtained from nucleation of a clear solution can be applied to the design of a continuous crystallizer.

In a heterogenous nucleation process, the nucleation rate should depend on both concentration and total crystal surface. In the presence of a large amount of crystals, the dependence on surface is a small effect as compared with the very strong and nonlinear dependence on supersaturation. Rumford and Bain (*14*) have investigated the nucleation of sodium chloride in a bed of large crystals and found that the dependence of the nucleation rate on supersaturation is very similar to that of homogenous nucleation. There is a pronounced measurable metastable supersaturation, (1.5 g. of salt/liter), below which nucleation is negligible and above which its rate increases rapidly and almost linearly with supersaturation. Close to the metastable region Equations (40) and (43) express therefore equally well the basic behavior of a nucleation process in a crystal magma. The advantage of using Equation (40) in a dynamic study is that it is smooth over the whole concentration range.

Larson (*12*) has suggested that the nucleation rate can be approximated by substituting c_s instead of c_m into Equation (43):

$$B = K(c - c_s)^m$$

This leads to a significant simplification of the mathematics, as now B can be expressed as a simple function of G. However, this approximation is only valid when $c_m - c_s$ is very small as compared with $c - c_s$. This might apply to precipitation of very insoluble compounds but is hardly a valid assumption for crystallizers which in most cases operate very closely to the metastable range (*14*).

STABILITY ANALYSIS

Clear Feed

The simplification of a clear feed, such as that $\varepsilon_0 = 1.0$ is imposed here. Many crystallizers operate in this manner since it alleviates the problem of preparing and metering seed nuclei into the system. This assumption means the following:

$$\psi = 0$$
$$\eta_n = 0 \qquad n = 0, 1, 2, \ldots$$
$$\varepsilon_0 = 1.0$$

which considerably simplified the relations previously derived.

The dimensionless groups arising in the set of linearized Equations (34) reduced to two, L_n and R_n, as defined in Equations (35) and (36) since the S_n are zero for the clear feed case. Before substituting the steady state relations (25) into (35) and (36) to determine L_n and R_n, we define a new group α:

$$\alpha = \left(\frac{\overline{\omega}}{V\,\overline{G}}\, r_0\right) = \frac{r_0}{\overline{\theta}\,\overline{G}} \tag{45}$$

This group is the ratio of the nuclei radius divided by the steady state growth, per draw-down time $v/\overline{\omega}$. Now substituting (25) and (45) and (36), we get

$$\begin{aligned} L_0 &= 0 \\ L_1 &= 1/(\alpha + 1) \\ L_2 &= 2(\alpha + 1)/(\alpha^2 + 2\alpha + 2) \\ L_3 &= 3(\alpha^2 + 2\alpha + 2)/(\alpha^3 + 3\alpha^2 + 6\alpha + 6) \end{aligned} \tag{46}$$

$$\begin{aligned} R_0 &= (1 - \overline{\varepsilon})/\overline{\varepsilon} \\ R_1 &= (1 - \overline{\varepsilon})\,\alpha/\overline{\varepsilon}(\alpha + 1) \\ R_2 &= (1 - \overline{\varepsilon})\,\alpha^2/\overline{\varepsilon}(\alpha^2 + 2\alpha + 2) \\ R_3 &= (1 - \overline{\varepsilon})\,\alpha^3/\overline{\varepsilon}(\alpha^3 + 3\alpha^2 + 6\alpha + 6) \end{aligned} \tag{47}$$

The groups L_n are functions of α only, while R_n are functions of both α and $\overline{\varepsilon}$. Values of L_n and R_n for the fourth and fifth moments can also be found without difficulty, when needed.

The closed set of linearized equations can now be written as

$$\begin{aligned}
&\frac{dZ_0'}{d\tau} + Z_0' + R_0 Z_3' - bY' = -p(\tau) \\
&\frac{dZ_1'}{d\tau} - L_1 Z_0' + Z_1' + R_1 Z_3' - [L_1 g + (1 - L_1)\, b]\, Y' = -p(\tau) \\
&\frac{dZ_2'}{d\tau} - L_2 Z_1' + Z_2' + R_2 Z_3' - [L_2 g + (1 - L_2)\, b]\, Y' = -p(\tau) \\
&\frac{dZ_3'}{d\tau} - L_3 Z_2' + [1 + R_3]\, Z_3' - [L_3 g + (1 - L_3)\, b]\, Y' = -p(\tau) \\
&\frac{dY'}{d\tau} + L_3 Z_2' - R_3 Z_3' + [L_3 g + (1 - L_3)\, b + 1]\, Y' = p(\tau) + q(\tau)
\end{aligned} \tag{48}$$

The eigenvalues of this set of equations can be found directly or the set of linear differential equations can be reduced to a set of algebraic equations by taking the Laplace transforms. In the latter case, the location of the poles of the transfer function would determine the stability of the system.

α, $\overline{\varepsilon}$, b, and g are the parameters of the system which determine the location of the poles of the transfer function. The results of applying the Routh-Horwitz criterion to this set of equations are presented in Figures 2 and 3. It is very surprising (Figure 2) that $\overline{\varepsilon}$ has such a negligible effect on the limits of stable operation.

It was found that for small values of α (which is most pertinent to crystallization), the ratio of b/g reached a limiting value with increasing values of g. For this reason

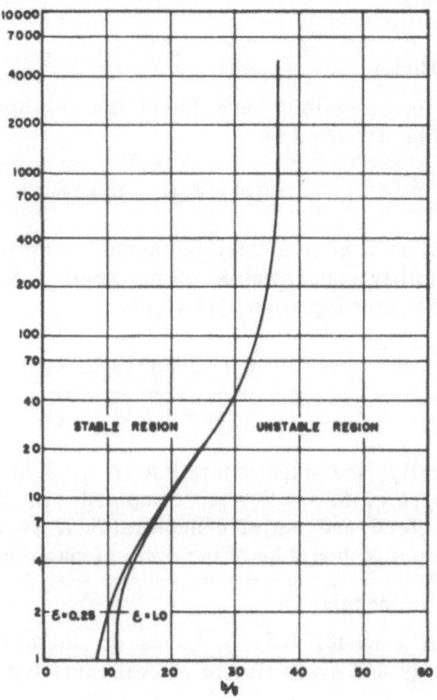

Fig. 2. Stability limits as a function of liquid volume fraction ε ($\alpha = 0.1$).

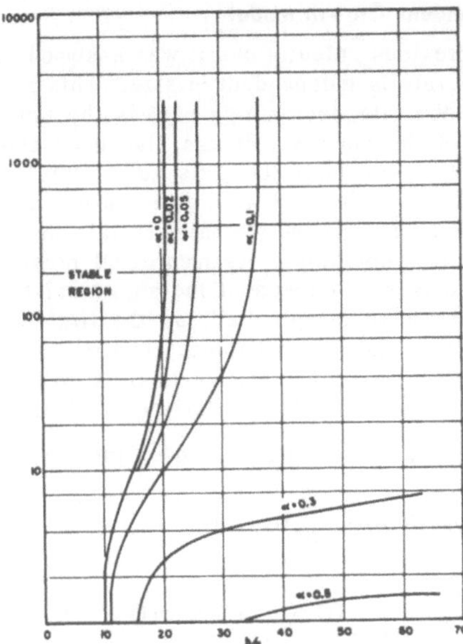

Fig. 3. Stability limits as a function of nuclei size parameter α ($\epsilon = 0.8$).

Figures 2 and 3 indicate g and b/g as the system parameters, rather than g and b. For high values of g, (which is expected), the stability depends on the single parameter b/g (critical) which is a unique function of α or the ratio of nuclei size to the average linear size increase of a crystal per draw-down time (Figure 3).

For most crystallization systems α will have values close to zero and in the past investigators (*2, 12, 15*) have taken α equal to zero. In some cases there is however a justification to consider values of α different from zero. Consider for example the case in which the growth rate is controlled by mass transfer to the surface. For particles which are large as compared with the microscale of turbulence (*16*) the mass transfer rate is independent of diameter. For small particles it is inversely proportional to particle size. To investigate the effect of the highly linearized growth rate of small particles on the dynamic behavior of the system, one can approximate the nonlinear dependence of G on r by a step function at which G is very large for $r < r_0$ and independent of r for $r > r_0$. This is equivalent to saying that the nuclei are really created at size r_0 which could be as large as 1 to 10 micra.

Seeded Feed

For the special case when the seed size is identical to the generated nuclei size (both distributions appearing as delta functions), the equations of the preceding section can be used to predict the effect of seed addition on the system's stability. The nucleation term $B(c)$ is expanded to include all nuclei entering the system as shown below.

$$B(c)_{\text{total}} = B(c)_{\text{nucleated}} + \frac{\psi\,\omega/V}{\varepsilon} \tag{49}$$

where ψ is defined to be the number of seeds per unit volume of feed so that $(\psi\,\omega/V/\varepsilon)$ is the seed addition rate per unit volume of crystallizer solution.

If we represent the steady state number of seeds as equal to ζ time the generated nuclei (bar indicates steady state values)

$$\frac{\overline{\psi}\,\overline{\omega}/V}{\overline{\varepsilon}} = \zeta \overline{B}(c)_{\text{nucleated}} \quad \zeta > 0 \tag{50}$$

so that

$$\overline{B}(c)_{\text{total}} = (\zeta + 1)\,\overline{B}(c)_{\text{nucleated}} \tag{51}$$

the sensitivity coefficient b/g for the seeded case will be less than the sensitivity coefficient for the unseeded case at identical supersaturations.

$$\left(\frac{b}{g}\right)_{\text{seeded}} = \frac{1}{1+\zeta}\left(\frac{b}{g}\right)_{\text{unseeded}} \tag{52}$$

Therefore at identical superation levels seed addition will increase the stability limits of a crystallizer while at the same time it reduces the weight mean particle size.

To determine the stability boundaries of the seeded system when the seed size is different from the nuclei size, it is necessary to return to the set of linearized Equations (34) and their associated dimensionless groups L_n, R_n, and S_n as given in Equations (35) to (37).

In order to evaluate these groups one can simplify matters by assuming that the seed distribution is a delta function at seed size r_s, such that

$$\eta_n = \int \psi(r)\,\delta(r - r_s)\,r^n\,dr = r_s^{\,n}\psi \tag{53}$$

Then two new steady state dimensionless groups are defined:

$$\beta = \overline{\omega} r_s/V\overline{G} = \frac{r_s}{\overline{\theta G}} \tag{54}$$

$$\zeta = \overline{\psi}(\overline{\omega}/V)/\overline{\varepsilon}\,\overline{B} = \overline{\psi}/\overline{\theta\,\varepsilon\,B} \tag{55}$$

The first of these β is similar to α defined in Equation (45) and is the ratio of the seed size to the steady state size increase per draw-down time. The second group ζ is the ratio of the seed addition rate per unit volume of crystallizer to the nuclei generation rate per unit volume of crystallizer. Substituting (45), (54), and (55) into the steady state moment relations (25), we simplify their representation as shown below:

$$\begin{aligned} \overline{\mu}_0 &= \overline{\varepsilon\beta\theta}(1+\zeta) \\ \overline{\mu}_1 &= \overline{\varepsilon\beta\theta}(\overline{\theta G})\,[(\alpha+1) + \zeta(\beta+1)] \\ \overline{\mu}_2 &= \overline{\varepsilon\beta\theta}(\overline{\theta G})^2\,[(\alpha^2+2\alpha+2) + \zeta(\beta^2+2\beta+2)] \quad \text{etc.} \end{aligned} \tag{56}$$

so that

$$\begin{aligned} L_n &= n\overline{\theta G}\,\frac{\overline{\mu}_{n-1}}{\overline{\mu}_n} = \text{funct. }(\alpha,\beta,\zeta) \\ R_n &= K\,\frac{\overline{\mu}_3}{\overline{\mu}_n}\,r_0^{\,n}\,\overline{\theta B} = \frac{1-\overline{\varepsilon}}{\overline{\mu}_n}\,r_0^{\,n}\,\overline{\theta B} = \text{funct. }(\alpha,\beta,\zeta,\varepsilon) \\ S_n &= \frac{\overline{\eta}_n}{\overline{\mu}_n} = \frac{r_s^{\,n}\overline{\psi}}{\overline{\mu}_n} = \text{funct. }(\alpha,\beta,\zeta) \end{aligned} \tag{57}$$

The behavior of the linearized equations describing the seeded case is determined by the parameters α, β, $\overline{\varepsilon}$, ζ, g,

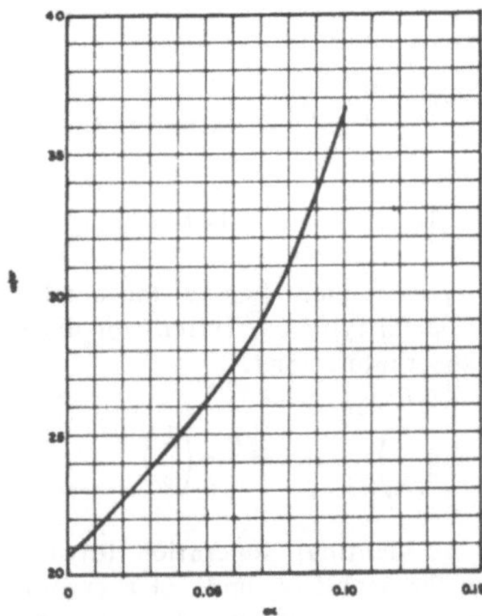

Fig. 4. Limiting values of b/g vs. α ($0 \leq \alpha \leq 0.1$).

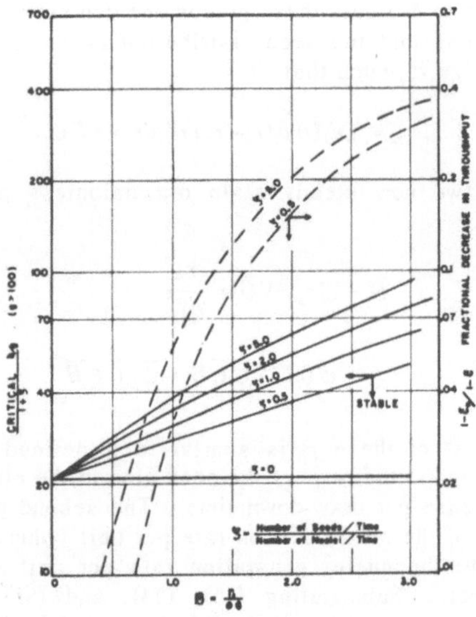

Fig. 5. Effect of seeding on stability and throughput.

and b; the two new groups β and ζ are added to characterize the seed size and quantity.

The stability boundaries for the seeded feed was investigated for a number of cases and the results are presented in Figure 4. These curves are for a nuclei size of zero ($\alpha = 0$) and a value of $g = 500$. The stability boundaries, as in the unseeded case, are relatively insensitive to variations in $\bar{\varepsilon}$ and g (when g is greater than 100). The stability limits increase with increasing seed number and seed size, both of which contribute to increasing surface area. However, larger seed sizes ($\beta > 1.0$) are not realistic since for a fixed feed rate or draw-down time the net throughput of product is diminished. Figure 5 also shows this relation by plotting the ratio of the crystal fraction of the feed over that of the product stream, against seed size at two levels of seed addition. The ordinate represents the loss of production capacity due to seeding.

Size-Dependent Growth Model

In the previous calculations it was assumed that the linear growth rate is independent of size. This assumption is correct if the rate determining step is the kinetic deposition rate of the surface. It has also been shown experimentally that for large crystals ($d >> 100\mu$) the overall mass transfer coefficient at high agitation rates is independent of size. For very small crystals this assumption probably does not hold. As mentioned previously if the growth rate is size dependent the analytical treatment becomes much more complicated, as the first four moments equations plus the solute balance equation are no longer a closed set of equations. However it is possible to estimate the nature of the effect of size dependent growth rates on stability by linearizing the growth function $\frac{dr}{dt}(r, c)$. If one can approximate $\frac{dr}{dt}(r, c)$ by

$$\frac{dr}{dt} = G(c)\,\phi(r) = G(c)\,(1 + ar) \tag{58}$$

then the set of Equations (11) and (20) will still give a self contained set.

When performing the calculations for this case it was assumed that the term $aG\theta$ was small, that is, of the order of ± 0.1 or less. The group $aG\theta$ is the relative size increase per draw-down time due to the size-dependent term ar, and the assumption that its value remains small is synonymous with the assumption that we are investigating a first-order correction of the size-independent ($\phi = 1$) growth model.

Figure 6 shows the effect of the size-dependent term on the stability contours for nonseeded operation with the nuclei size (α) equal to zero. It is seen that a model which predicts increasing growth rate with increasing size raises the system's stability limits. Based on the previous re-

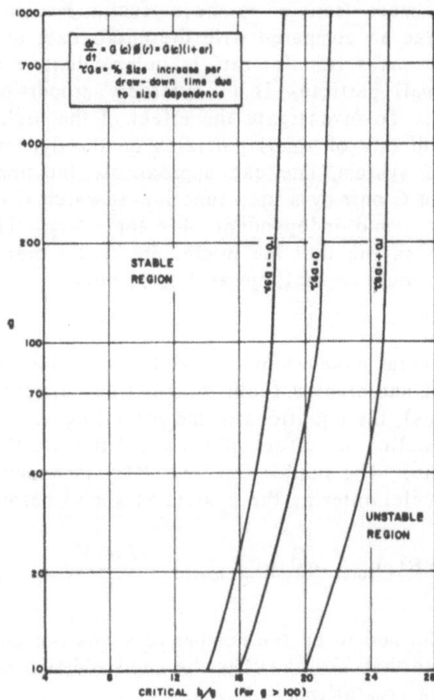

Fig. 6. Effect of linear $\phi(r)$ on stability.

sults this is to be expected, since it should lead to a shorter time lag between the appearance of increased nuclei and increased surface area in the system.

Operation with a Fines Trap

In order to increase crystal product size many crystallization systems utilize a nuclei trap. This usually consists of withdrawing solution from some part of the vessel in such a manner that only crystals below a certain size will be elutriated with the solution. This stream is heated to redissolve the solids and then returned to the crystallizer body. Most often this operation is carried on continuously rather than on the basis of some feedback signal. The question we wanted to answer at present is whether this continuous nuclei removal operation would increase or decrease the system's stability.

In order to dissolve the nuclei they must be trapped very early, before they have a chance to grow, say within $\frac{1}{8}$ to $\frac{1}{4}$ of a draw-down time θ. Figure 7 indicates that over this time period, even for cycling behavior, the variation in supersaturation or growth rate $G(c)$ is relatively small. We therefore assume a quasi steady state condition such that the fraction of new nuclei being generated at a specific time which survive the trap can be related to the instantaneous growth rate.

The residence time distribution function $F(t)$ is the fraction of particles which reside in a vessel for a period of time less than t. For a well-stirred vessel this function is

$$F(t) = 1 - e^{-t/\theta'} \tag{59}$$

where θ' is a draw-down time, and for this analysis is based on the volumetric flow to the nuclei trap. If the critical size for removal is denoted as r_c, we can denote the fraction of newly formed nuclei lost to the trap as

$$\text{Fraction lost} = 1 - e^{-\frac{r_c}{G(c)\theta'}} \tag{60}$$

since we have made the quasi steady state assumption over this small time period. Therefore the net nucleation rate for the crystallizer and associated nuclei trap is

$$\text{Net nucleation} = B(c)\, e^{-\frac{r_c}{G(c)\theta'}} \tag{61}$$

To determine the stability of this system we can utilize the previously derived analysis presented in the beginning of this section. All we need do is use Equation (61) as the nucleation function in determining the sensitivity parameter b/g. We see then that for this case

$$\frac{b}{g} = \frac{\overline{G}}{\overline{B}} \frac{\frac{dB}{dc}(\overline{c})}{\frac{dG}{dc}(\overline{c})} + \frac{r_c}{\theta'\overline{G}} \tag{62}$$

whereas for operation without the fines trap

$$\frac{b}{g} = \frac{\overline{G}}{\overline{B}} \frac{\frac{dB}{dc}(\overline{c})}{\frac{dG}{dc}(\overline{c})} \tag{63}$$

all the terms being evaluated at the steady state.

At a specified solute concentration the value of Equation (62) is always larger than Equation (63). Alternately, from Equation (42) we see that the critical supersaturation level, beneath which cycling occurs, is higher when a nuclei trap is added to the system. Since the steady state solute concentration is slightly increased when a nuclei trap is added to the system, a specification must be set for the two modes of operation before a comparison can be made as to their relative stability.

A meaningful comparison between operation with and without a fines trap is to specify that with the same feed composition, the same product should be produced (for example, the same weight mean particle size). Since the supersaturation is greater with the operation of a fines trap than without, the production rate will also be greater for this case. It is shown in Appendix A that for this case, the addition of a fines trap to a crystallizer to produce the same product at a higher capacity will not reduce the stability of the system, and in some cases would improve it. This conclusion has been reached just for the case where the trap is continuously operated and it should be noted that the results could be quite different when the trap is operated on an on-off basis under operator control.

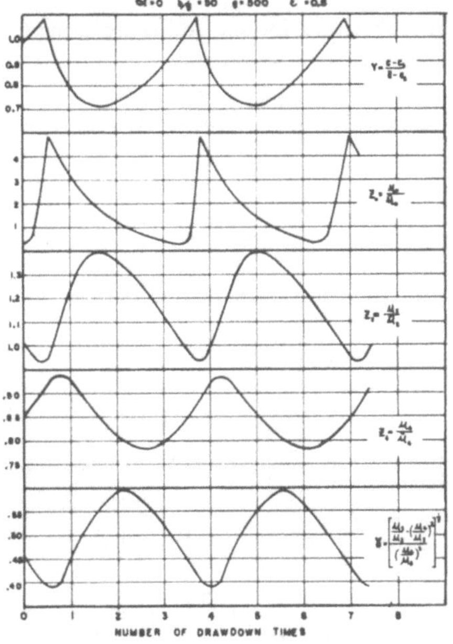

Fig. 7. Typical limit cycle behavior. $\alpha = 0$, $b/g = 50$, $g = 500$, $\epsilon = 0.8$.

NONLINEAR SOLUTIONS

Clear Feed

In the previous sections, the stability of the system was analyzed by linearization of the equations. This allows one to predict the general effect of the physical parameters on the system and is especially useful when investigating the possibility of stabilizing such a crystallizer by changing the crystallization conditions, seeding, or some feedback control. However in order to understand the complete behavior of the system it is illuminating to solve the complete nonlinear set of equations which can only be done numerically (or on an analog computer).

The starting point of the following analysis is the system of Equations (22), without the seed terms.

$$\frac{d\mu_n}{dt} = nG\mu_{n-1} + (1 - K\mu_3) B r_0^n - \frac{\mu_n}{\theta} \quad n = 0, 1, 2, 3 \ldots \quad (64)$$

$$(1 - K\mu_3) \frac{dc}{dt} = \frac{c_0 - c}{\theta} - (\rho - c)(3K\mu_2 + (1 - K\mu_3) BK r_0^3) \quad (65)$$

To simplify the study of dynamic behavior of this type of system, the variables are normalized about the steady state variables corresponding to the feed concentration c_0 and rate ω. If the feed concentration and rate vary periodically, their mean values can be used as reference points.

Instead of defining the concentration variable as being the ratio of absolute concentration divided by the steady state value, which would be extremely close to 1 at all times, the normalized supersaturation will be followed in time.

The following dimensionless groups are defined:

$$Z_n = \frac{\mu_n}{\overline{\mu}_n} \quad n = 0, 1, 2, \ldots$$

$$Y = (c - c_s)/(\overline{c} - c_s) \quad (66)$$

where c_s is constant since the system is isothermal.

Since the feed concentration and rate can vary with time, we define

$$j(\tau) = \frac{\omega(\tau)}{\overline{\omega}}$$

$$h(\tau) = \frac{c_0(\tau)}{\overline{c}_0} \quad (67)$$

$$\tau = t\,\overline{\omega}/V = t/\theta$$

where $\overline{c}_0$ and $\overline{\omega}$ are the average values of these inputs. If the feed conditions are constant $j(\tau)$ and $h(\tau)$ are 1.

If these groups are substituted into (64) and (65), along with the dimensionless steady state groups defined in (35) and (36), we get

$$\frac{dZ_n}{d\tau} = L_n \left(\frac{G}{\overline{G}}\right) Z_{n-1} + \left(\frac{B}{\overline{B}}\right) \frac{R_n}{1 - \overline{\varepsilon}} - \left(\frac{B}{\overline{B}}\right) R_n Z_3 - Z_n j(\tau) \quad n = 0, 1, 2, \ldots$$

and recognizing that $dc/dt = d(c - c_s)/dt$, we get

$$(1 - (1 - \overline{\varepsilon}) Z_3) \frac{dY}{d\tau} = j(\tau) \left[h(\tau) \frac{\overline{c}_0 - c_s}{\overline{c} - c_s} - Y \right] + j(\tau)\,[h(\tau) - 1] \frac{c_s}{\overline{c} - c_s} - \left[\frac{\rho - c_s}{\overline{c} - c_s} - Y \right] \times$$

$$\left[L_3 \left(\frac{G}{\overline{G}}\right) (1 - \overline{\varepsilon}) Z_2 + (1 - (1 - \overline{\varepsilon}) Z_3) R_3 \left(\frac{B}{\overline{B}}\right) \right] \quad (68)$$

The two concentration groups in Equation (68)

$$\frac{\overline{c}_0 - c_s}{\overline{c} - c_s} \quad \text{and} \quad \frac{\rho - c_s}{\overline{c} - c_s}$$

can be rearranged in terms of groups previously defined in the stability study. When we take the kinetic growth model to be linear in supersaturation

$$G = K_1 (c - c_s) \quad (38)$$

then as before

$$g = \frac{1}{\overline{\varepsilon}} \frac{\overline{c}_0 - \overline{c}}{\overline{c} - c_s} \quad (39)$$

and

$$\frac{\overline{c}_0 - c_s}{\overline{c} - c_s} = \frac{c_0 - \overline{c}}{\overline{c} - c_s} + 1 = 1 + \overline{\varepsilon} g \quad (69)$$

and substituting Equation (26) (with $\varepsilon_0 = 1.0$) into the second group, we get

$$\frac{\rho - c_s}{\overline{c} - c_s} = \frac{\rho - \overline{c}}{\overline{c} - c_s} + 1 = \frac{c_0 - \overline{c}}{\overline{c} - c_s} \frac{1}{1 - \overline{\varepsilon}} + 1 = \frac{\overline{\varepsilon}}{1 - \overline{\varepsilon}} g + 1 \quad (70)$$

and

$$\frac{G}{\overline{G}} = \frac{K_1 (c - c_s)}{K_1 (\overline{c} - c_s)} = Y \quad (71)$$

Also, by assuming a Volmer nucleation model, and remembering we are assuming that c/c_s is close to 1, we can write the nucleation rate as

$$B = K_2\, e^{-\frac{K_3}{[\ln c/c_s]^2}} \approx K_2\, e^{-\frac{K_3}{(c/c_s - 1)^2}} \quad (40)$$

and in conjunction with the linear growth model

$$\frac{b}{g} = \frac{2K_3}{\left(\frac{\overline{c}}{c_s} - 1\right)^2} \quad (42)$$

so we can substitute for the following groups $(c_s/\overline{c} - c_s)$ and $B/\overline{B}$

$$\frac{c_s}{\overline{c} - c_s} = \left[\frac{b/g}{2K_3}\right]^{1/2} \quad (72)$$

$$\frac{B}{\overline{B}} = e^{\frac{1}{2}\frac{b}{g}\left(1 - \frac{1}{Y^2}\right)} \quad (73)$$

and substitute (69) to (73) back into (68)

$$\frac{dZ_n}{d\tau} = Y L_n Z_{n-1} + e^{\frac{1}{2}\frac{b}{g}\left(1 - \frac{1}{Y^2}\right)} \left[\frac{R_n}{1 - \overline{\varepsilon}} - R_n Z_3 \right] - Z_n j(\tau) \quad n = 0, 1, 2, 3, \ldots \quad (74)$$

$$(1 - (1 - \overline{\varepsilon}) Z_3) \frac{dY}{d\tau} [h(\tau)(\overline{\varepsilon} g + 1) - Y] + j(\tau)\,[h(\tau) - 1] \left[\frac{b}{g} \Big/ 2K_3\right]^{1/2} - \left[\frac{\overline{\varepsilon}}{1 - \overline{\varepsilon}} g + 1 - Y \right] \times \left[L_3 Y (1 - \overline{\varepsilon}) Z_2 + (1 - (1 - \overline{\varepsilon}) Z_3) R_3\, e^{\frac{1}{2}\frac{b}{g}\left(1 - \frac{1}{Y^2}\right)} \right] \quad (75)$$

This set of equations describes the behavior of the system relative to its steady state in terms of the parameters used in the linear stability analysis (remembering L_n is a function of α and R_n is a function of α and $\overline{\varepsilon}$). Therefore with the linear growth model, the modified Volmer nucleation model, and the assumption of steady state feed conditions ($j(\tau)$ and $h(\tau)$ equal to 1.0), we see the relative dynamic behavior is determined *exactly* by the same variables which determine stability (α, g, $\overline{\varepsilon}$, b/g), while for time varying feed conditions, the constant K_3 must also be known.

Alternately, if the metastable nucleation model were used, the term $(B/\overline{B})$ could be arranged

$$\left(\frac{B}{\overline{\overline{B}}}\right) = \left(\frac{Y - 1 + \dfrac{m}{b/g}}{m \Big/ \dfrac{b}{g}}\right)^m \qquad Y > 1 \tag{76}$$

indicating that in comparison with the Volmer model, the behavior of the system depends on an additional parameter m, which does not appear in the linear stability analysis. The advantage of using the Volmer model which, in most cases, fits the data equally well is that the number of dimensionless groups is reduced. We will later show that in this case the dimensionless behavior of most crystallizers can be characterized by a single dimensionless group b/g which is relatively insensitive to the kinetic model chosen.

Equation (68), along with appropriate nucleation models and an initial perturbation, were solved numerically with the City College IBM 7040 computer.

In all cases in which the system parameters indicated linear stability, initial perturbations were indeed damped out. Furthermore, it was found that in all cases for which the system parameters indicated linear instability, limit cycles developed for any size perturbation. This indicates that the regions mapped out by the linearized stability analysis are representative of the behavior of the actual system.

Limit cycle behavior is caused by operation at a point where the nucleation function is very sensitive (high b/g) which causes the system to give rise to a shower of nuclei when the equilibrium concentration has been exceeded. These nuclei then grow, supplying so much area for growth that the solute concentration decreases below its equilibrium value. This in turn causes the generation of much less than the equilibrium supply of nuclei. Continued withdrawal of crystals from the system reduces the available area for growth and this causes the solute concentration to rise, which again results in a shower of nuclei and repitition of the cycle. Figure 7 demonstrates just this behavior for a typical limit cycle.

The remainder of the work presented in this section focused on the properties of these limit cycles.

The majority of the numerical work done with the nonlinear system Equations (74) and (75) utilized the modified Volmer nucleation model. For this choice of kinetic model [which includes a linear growth model (38)], the nonlinear system is completely defined by the dimensionless groups used in the linearized stability analysis.

Other interesting properties were predicted by Figures 2 and 3 which required corroboration by studying the behavior of the nonlinear system. Specifically, for values of g greater than about 100, the linearized stability indicated that the critical value of b/g was independent of g and that the critical value of b/g was independent of $\overline{\varepsilon}$. Numerous results did indeed indicate for moderately high values of g that the normalized dynamic behavior of the system is independent of g and $\overline{\varepsilon}$.

Since the assumption of $g > 10^2$ is realized for most crystallization systems, the dynamic behavior of the system relative to its predicted steady state becomes a function of only two variables (b/g and α). Then once α has been specified, the value of b/g is the only parameter of importance in determining the limit cycle behavior of the systems.

It was mentioned previously that for a different choice of the nucleation model such as the metastable model, an additional dimensionless group m was introduced. One would, however, expect, from the results based on the modified Volmer model, that the effect of changing m, while keeping b/g constant, should be negligible. This is due to the fact that both nucleating functions are similar near the so-called metastable region and it would be hard to differentiate between them on the basis of experimental data.

The effect of m on the limit cycle was evaluated by determining numerically how the parameters of the limit cycle varied as a function of m. For values of g and α for which the critical value of b/g is independent of g and $\overline{\varepsilon}$, the effect of changing m on the parameters of the limit cycle was negligible.

Curves describing the behavior of the limit cycles are presented in Figures 8, 9, and 10 for α values of 0.0 and

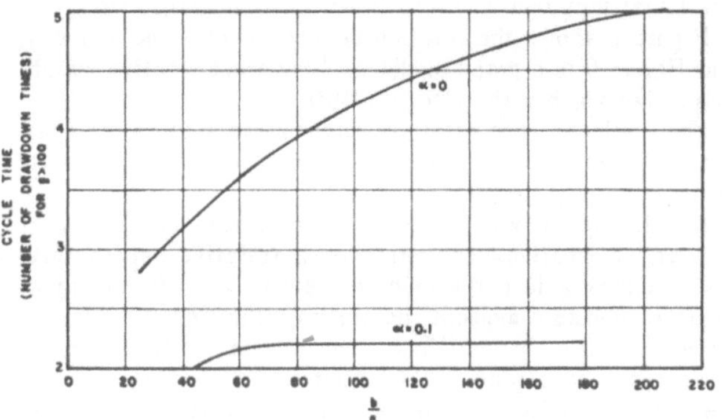

Fig. 8. Cycle time vs. b/g for $g > 100$.

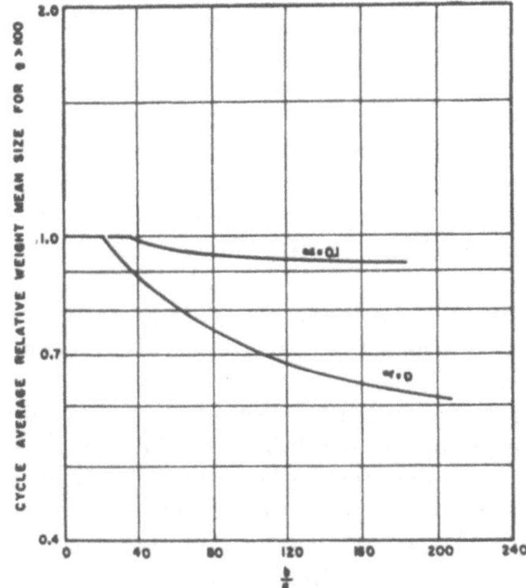

Fig. 9. Cycle average relative weight mean size (Z_4/Z_3) vs. b/g.

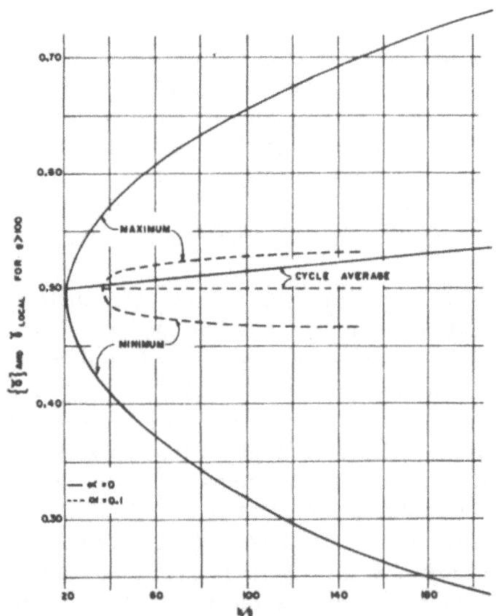

Fig. 10. Cycle average and local coefficient of variation of the weight distribution vs. b/g.

0.1. Actually, the nuclei formed in most systems are so small that they can be taken to be approximately zero.

Figure 8 shows the relation between cycle time and b/g, and Figure 9 is a graph of the cycle average relative weight mean size vs. b/g (both for $g > 10^2$).

The cycle average relative weight mean size is the average of Z_4/Z_3 over a cycle and is indicative of a composite of continuously sampled material.

As the sensitivity of the system increases (b/g gets greater), a small excess value of Y (relative supersaturation) causes a larger amount of new nuclei (the maximum value of the zero moment increases), requiring a longer interval of time to withdraw the crystals before another shower can take place. This increase of b/g also raises the average surface area present over a cycle, thereby lowering the weight mean product size. Both these effects are shown in Figures 8 and 9.

The effect of increasing the size of α is to damp the amplitude of the zero moment. This is due to the increased size of the nuclei which when formed rapidly reduce the supersaturation, causing less nuclei to be formed during a disturbance than would be for a system with an α of zero. This in turn means that the cycle time will be shorter and the relative weight mean size larger than for a system with a negligible α. This is also shown by Figures 8 and 9. However for α to have an appreciable effect it would have to be much larger than normally encountered during crystallization.

Figure 10 relates the local and cycle average (composite) coefficient of variation of the weight distribution to b/g. This indication of the "tightness" of the distribution is equal to 0.5 for a crystallization system operating in the steady state (both for $\alpha = 0.0$ and 0.1). It is shown in Appendix B that the composite value of γ can never be better than the steady state value. The actual composite value of γ when $\alpha = 0.0$, is only slightly greater than 0.5, while when $\alpha = 0.1$ the difference from 0.5 could not even be noticed. Therefore on a time-averaged basis the non-stable system gives nearly as good a product as does a stable system.

Stabilization by Seeding

There is much practical experience in industry which indicates that adding seed to an oscillating crystallizer will stabilize its operation. The results of the last section seem to corroborate this.

If one had an oscillating system at hand it would be interesting to learn just how much seed of what size would be necessary to stop the oscillations and how the resulting product distribution would change. We have performed this analysis for one case where the initial steady state parameters were $\alpha = 0$, $\overline{\varepsilon} = 0.8$, $g = 500$, $b/g = 50$. These parameters represent an unstable system as can be seen from Figure 2. From Figure 9 it is seen that the resulting product distribution has a cycle average weight mean size of only 85% of that predicted by the steady state equations. Figure 10 shows that substantial local fluctuations in the coefficient of variation will occur while the resulting average is only slightly higher than that of steady state operation.

To perform the calculation we assumed a linear growth model and modifier Volmer nucleation model such that the group b/g is equivalent to that given in Equation (42). The volumetric feed rate remained constant and for each seed size a trial and error calculation was performed to determine how much seed was necessary to stabilize the system. When this was found the weight mean size relative to that predicted by the initial, but unstable, steady state could be calculated. Other properties such as the coefficient of variation and the feed voidage, the latter being an indication of lost productivity as it becomes less than 1, were also calculated.

The results of these calculations are shown in Figure 11. It can be seen for values of β below 1 there is an imperceptibly small difference in the resulting weight mean size, variance, or productivity as compared with the initial, but unattainable, steady state values. The reason for this is that the final steady state has a lower supersaturation than the original unseeded steady state had (final b/g equals 54, original equalled 50), which results in a lower nucleation rate, thus tending to offset the effect of seed addition.

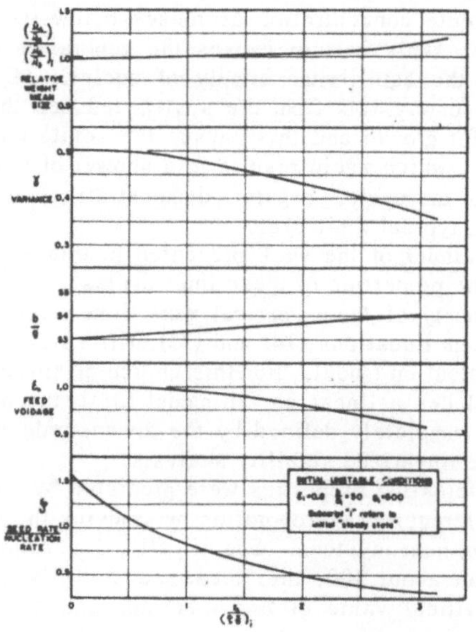

Fig. 11. Effect of seed size stabilization.

It appears therefore that seed addition is an excellent solution to end cycling performance by both stabilizing the operation and increasing the weight mean size of the product without any significant loss of production.

The results of this study can also be used to analyze the performance of an imperfectly mixed crystallizer. For example, poor mixing at the feed inlet would lead to local high supersaturation levels and local high nucleation rates while the rest of the crystallizer is well mixed. This kind of system is analogous to a seeded feed into a well-mixed crystallizer, indicating why many commercial installations are free of the problem of cycling.

DISCUSSION OF RESULTS

As mentioned in the introduction, it is well established that continuous crystallizers exhibit cycling behavior. The occurrence of cycling behavior is probably much more frequent than is realized. As shown in the analysis the cycle period is very long, about three to five times the solids' draw-down times. Such long cycles might therefore involve several operator shifts and the cycle itself might be strongly perturbed or masked by changes introduced by the operator. Depending on the control strategy used by the operator, this might enhance instability or just cause an increased variability in the size distribution.

The data available in the literature from which one can calculate values of b/g for a particular system are extremely sparse. However, the two sources which we found both agree with the results of this study.

1. Robinson and Roberts (*13*) fitted a metastable model to an ammonium sulfate system they were operating in their plant. The plant crystallizer was well stirred and did not classify the product. Utilizing Equation (44) and their model, we were able to determine that b/g for this system was greater than 20. Instability and long-term transients were reported for this system, and to calculate the steady state particle distribution an average of sixty-one samples was taken over a period of days.

According to the results of this study, the stability limit for this type of operation is a b/g of about 21 and in all likelihood the instability reported for this system was inherent to the system rather than due to the operators. The fact that they obtained a steady state particle balance which fitted steady state theory only by averaging their results over a period of days is also predicted by this study, as shown in Figure 10.

2. The study of Bransom (*21*) with a well-stirred isothermal salting out crystallizer provides the best data available in the literature. He studied a cyclonite system with the supersaturation provided by the addition of alcohol. By fitting a modified Volmer model to his data, we found that b/g never exceeded a value of 4.0 for the conditions of his study. Consequently, no cycling problem was reported and a particle size distribution was attained which fit steady state theory without averaging. Residence times were very low for this study (about 15 min.) and the product crystal was therefore very small. Calculations indicate that if the residence time were increased to about 1 hr. (which would increase the product size), the stability limits would be reached and any attempt to grow still larger crystals would result in cycling.

SUMMARY AND CONCLUSIONS

The analytical treatment given in this paper provides a basis of understanding the general nature of these phenomena. It is seen that the dominating source of this unstable behavior is the strongly nonlinear nature of the dependence of nucleation rate on supersaturation which, in an oversimplified form, is sometimes described by assuming the existence of a metastable region in which no nucleation occurs. The larger the particle size the lower must be the nucleation rate, and therefore for large crystals one has to operate at supersaturations close to the so-called metastable region. A small upset at these conditions will cause a temporary increase in nucleation rate. Now as the excess amount of nuclei grows the total area increases and the supersaturation decreases, reducing the nucleation rate. However, this effect occurs with a considerable time delay as the new nuclei have no appreciable surface for a considerable time span. Therefore, before the stabilizing action occurs a large number of nuclei might be formed which later will reduce the supersaturation so much that the solution becomes metastable and nucleation practically ceases. At some later time the total surface starts to decrease due to the removal of crystals from the crystallizer in the outflow and the supersaturation starts to increase again.

This leads therefore to the occurrence of limit cycles whose specific properties were described in the analytical part and can be summarized as follows.

Both the tendency to instability and the relative amplitude of the limit cycles increase with increasing b/g and therefore with increasing particle size, at otherwise constant conditions: (1) The free volume or the magma density has little effect on stability limits. (2) The cycle period is comparatively long (three to five times the draw-down time). (3) The instabilities tend to decrease the average particle size and increase the total variance. However, if an individual sample is withdrawn, chances are high that its size distribution will be more uniform than for a Poisson distribution. (4) Correct feeding of the seed will tend to stabilize the system and increase particle size. (5) The total crystal mass will in most cases remain almost constant.

In all the analysis it was assumed that the crystallizer is an ideally stirred tank. It should be pointed out in all fairness that for many crystallizers this is not a good approximation. In this model any incoming feed will become immediately dispersed. In reality this takes a finite mixing time varying from a fraction of a second to several seconds, depending on the size of the vessel (mixing time increases with vessel size (*16*)). With the use of such a model for a regular first- or second-order chemical reaction with a residence time large as compared with the mixing time, the error introduced by neglecting this short initial mixing period is often negligible. In any phenomena involving nucleation the increase in rate of nucleation during this period might be so large that almost all of it will occur during that stage.

Therefore one has to be very careful in using such a model in crystallization. In particular it is doubtful if one can calculate nucleation rates from the overall performance of such crystallizers, as has been suggested.

The deviation of an actually mixed crystallizer from that of an ideally mixed one will therefore depend both on the nature of the crystallizing solution and on the initial mixing time and therefore on the size of the vessel. The latter factor explains some of the special difficulties encountered in the scale-up of crystallizers.

This deviation from ideal mixing will have a similar effect as that of seeding and will have some stabilizing action. However, even in a case where there are considerable deviations between the actual behavior of the crystallizer and the ideally mixed theoretical model, the present analysis should still present the correct trends and provide an understanding of the phenomena leading to the cyclic behavior of crystallizers.

ACKNOWLEDGMENT

The writers are indebted to their friends at the American Cyanamid Company, and especially to H. M. Hulburt (now at Northwestern University), for their help with the early phases of this work.

The work reported here was supported in part under National Science Foundation Grant No. GK 943.

NOTATION

a = constant in size-dependent growth model, Equation (39)
b = dimensionless nucleation sensitivity group
B = nucleation rate per volume of solution
c = solute concentration in the crystallizer
c_0 = solute concentration in the feed
c_m = metastable concentration
c_s = solubility concentration of the solute
$f(r, t)\,dr$ = number of crystals per unit volume having radii in the range $r, r + dr$ at time τ
$F(t)$ = fraction of particles residing in a vessel for a period of less than t
g = dimensionless growth sensitivity group
G = crystal growth rate dependence on concentration
$h(\tau)$ = normalized feed concentration
$j(\tau)$ = normalized feed rate
K = crystal shape factor
K_n = constants
L_n = steady state dimensionless groups dependent on α
m = exponent in metastable nucleation function
$p(\tau)$ = dimensionless feed rate for stability analysis
$q(\tau)$ = dimensionless feed concentration for stability analysis
r = characteristic radius of the crystal
r_c = nuclei trap cutoff size
r_0 = nuclei characteristic radius
r_s = seed size
r_{wm} = mean particle size with regard to the weight distribution
R_n = steady state dimensionless groups dependent on α and ε
t = time
$\overline{v}$ = partial molar volume of solute
V = crystallizer working volume
$W(r)\,dr$ = crystal weight per volume having radii in the range $r, r + dr$
Y = normalized supersaturation
Y' = dimensionless concentration perturbation
Zn = normalized moments
Zn' = dimensionless moment perturbations

Greek Letters

α = dimension group as defined in Equation (25), $r_0/\tau G$
β = dimension group as defined in Equation (37), $r_s/\tau G$
γ = coefficient of variation
ε = fractional volume of solution
ε_0 = fractional volume of solution in feed
ζ = ratio of seed rate to nucleation rate
η_n = n^{th} moment of ψ
θ = crystallization draw-down time = V/w
θ' = draw-down time based on feed to the nuclei trap
μ_n = n^{th} moment of $f(r)$
ρ = crystal density
τ = dimensionless time
ϕ = crystal growth rate dependence on r
$\psi(r, t)\,dr$ = number of crystals per volume of feed having radii in the range $r, r + dr$, at time t
ω = volumetric feed and/or withdrawal rate

Subscripts

0 = properties of feed
1 = properties of take off

Superscripts

— = steady state values
$'$ = perturbation from steady state
$\{\ \}$ = cycle average
$\langle\ \rangle$ = integral over r weighted by f

LITERATURE CITED

1. Aris, Rutherford, and N. R. Amundson, *Chem. Eng. Sci.*, **7**, Pt. I-3, No. 3, 121 (1958).
2. Bransom, S. H., *Brit. Chem. Eng.*, **12**, 838 (1960).
3. ———, et al., *Discussions Faraday Soc.*, **5**, 83 (1949).
4. Cannon, K. J., and N. R. Amundson, *AIChE J.*, **9**, 297 (1963).
5. Crocco, N., and S. Cheng, "Rocket Instability," Butterworths, London (1956).
6. Finn, R. K., and R. E. Wilson, *Agr. Food Chem.*, **2**, No. 2, 66 (1954).
7. Han, C. D., and Reuel Shinnar, paper presented at the 61st National AIChE Meeting, Houston, Texas (February, 1967).
8. Hulburt, H. M., and Stanley Katz, *Chem. Eng. Sci.*, **19**, 555 (1964).
9. McCabe, W. L., and R. P. Stevens, *Chem. Eng. Progr.*, **47**, No. 4, 168 (1951).
10. Miller, P., and W. C. Saeman, *ibid.*, **43**, No. 12, 667 (1947).
11. Murray, D. C., and M. A. Larson, *AIChE J.*, **11**, No. 4, 728 (1965).
12. Randolph, A. D., and M. A. Larson, *ibid.*, **8**, No. 5, 639 (1962).
13. Robinson, J. N., and J. E. Roberts, *Can. J. Chem. Eng.*, **10**, 5 (Oct., 1957).
14. Rumford, F., and J. Bain, *Trans. Inst. Chem. Engrs.*, **38**, 10 (1960).
15. Saeman, W. C., *AIChE J.*, **2**, No. 1, 107 (1956).
16. Shinnar, Reuel, *J. Fluid Mech.*, **10**, Pt. 2, 259–275 (1961).
17. Tanimoto, A., K. Kobayashi, and S. Fujita, *Intern. Chem. Eng.*, **4**, No. 1, 153 (1964).
18. Thomas, W. M., and W. C. Mallison, *Petrol. Refiner*, No. 5, 211 (1961).
19. Van Hook, A., "Crystallization: Theory and Practice," ACS Monograph 152, p. 94, Reinhold, New York (1961).
20. *Ibid.*, p. 13.
21. Worden, R. B., and N. R. Amundson, *Chem. Eng. Sci.*, **17**, 725 (1962).

APPENDIX A.

Operation with and without a fines trap to produce the same product a comparison of stability

In the following let all parameters related to operation with a fines trap be denoted by a subscript f. We can write the nucleation rates for the two modes of operation as follows, using Equations (40) and (61), and assuming c/c_s is close to 1:

$$B_f = K_2\, e^{-\frac{K_3}{\left(\frac{c_f}{c_s}\right)^2}}\, e^{-\frac{r_c}{\theta' G_f}} \tag{1A}$$

$$B = K_2 e^{-\frac{K_3}{\left(\frac{c}{c_s}-1\right)^2}} \qquad (2A)$$

Assuming a linear growth model [Equation (38)], we can evaluate the sensitivity parameter b/g for the two modes of operation as described in Equations (62) and (63):

$$\left(\frac{b}{g}\right)_f = \frac{2K_3 c_s^2}{(c_f - c_s)^2} + \frac{r_c}{\theta' G_f} \qquad (3A)$$

$$\left(\frac{b}{g}\right) = \frac{2K_3 c_s^2}{(c - c_s)^2} \qquad (4A)$$

In order for the product to be the same for the two modes of operation, we require the weight mean size (μ_4/μ_3) to be the same.

From Equations (25) (unseeded), this yields

$$\theta_f G_f = \theta G \qquad (5A)$$

The similarity condition also requires the nucleation rate to be proportional to the feed rate or inversely proportional to the residence time so that

$$\frac{B_f}{B} = \frac{\theta}{\theta_f} \qquad (6A)$$

and combining (5A) and (6A)

$$\frac{B_f}{G_f} = \frac{B}{G} \qquad (7A)$$

Now ratioing $(b/g)_f$ and (b/g) from Equations (3A) and (4A), we get

$$\frac{(b/g)_f}{(b/g)} = \left(\frac{c - c_s}{c_f - c_s}\right)^2 + \frac{r_c/\theta' G_f}{(b/g)} \qquad (8A)$$

From Equation (38) it can be seen that

$$\left(\frac{c - c_s}{c_f - c_s}\right)^2 = \left(\frac{G}{G_f}\right)^2 \qquad (9A)$$

So that substitution of Equation (9A) and then (7A) into (8A) yields

$$\frac{(b/g)_f}{(b/g)} = \left(\frac{B}{B_f}\right)^2 + \frac{r_c/\theta' G_f}{(b/g)} \qquad (10A)$$

Substitution of (1A) and (2A) into (10A) results in

$$\frac{(b/g)_f}{(b/g)} = e^{-\frac{b}{g}\left[\frac{(b/g)_f}{(b/g)} + \frac{r_c/\theta' G_f}{(b/g)} - 1\right]} + \frac{r_c/\theta' G_f}{(b/g)} \qquad (11A)$$

We now wish to know if, with realistic values for $r_c/\theta' G_f$ and (b/g), the ratio $(b/g)_f/(b/g)$ will depart greatly from 1. Values of this ratio greater than 1 indicate a decreasing stability, while values less than 1 indicate the opposite.

Values for $r_c/\theta' G_f$ of 1.0 and 2.38 refer to 67 and 91% fines removal, respectively. With the assumption that the initial system is at the stability borderline (b/g near 20), the term $(r_c/\theta' G_f)/(b/g)$ would have a value in the range of 0.05 to 0.10 for the fines removal rates indicated above. With these values Equation (11A) can be solved, and it is found that the ratio $(b/g)_f/(b/g)$ is slightly smaller than 1, indicating that the addition of a fines trap will not itself cause instability but would actually improve the stability of the system.

APPENDIX B.

Relation of the cycle average coefficient of variation to the steady state value

This relation is derived for the case when $\alpha = 0$ and the feed is clear. For a crystallizer, the supersaturation is so small that μ_3 or ε is taken as a constant.

Steady State

$$\gamma^2 = \frac{\frac{\mu_5}{\mu_3} - \left(\frac{\mu_4}{\mu_3}\right)^2}{\left(\frac{\mu_4}{\mu_3}\right)^2} = \frac{\mu_3 \mu_5}{\mu_4^2} - 1 \qquad (1B)$$

Substitution of the relations of (25) for the steady state moments gives

$$\gamma^2 = \frac{5}{4} - 1 = \frac{1}{4}$$

$$\gamma = 0.5$$

Calculation will show that even for values of $\alpha = 0.1$, the steady state coefficient of variation will still be 0.5.

Limit Cycle Average

The average of a moment taken over a limit cycle of time (θ) can be written as

$$\{\mu_n\} = \frac{1}{\theta}\int_t^{t+\theta} \mu_n \, dt$$

In order to get a composite value for the coefficient of variation over the limit cycle, the cycle average variance should be evaluated with respect to the cycle average weight mean size.

$$\{\gamma^2\} = \frac{\frac{\{\mu_5\}}{\mu_3} - \frac{\{\mu_4\}^2}{\mu_3^2}}{\frac{\{\mu_4\}^2}{\mu_3^2}} \qquad (2B)$$

When $r_0 = 0$ and the feed is clear, the moment equations for μ_4 and μ_5 can be written from Equation (12) as

$$\frac{d\mu_4}{dt} = 4G\mu_3 - \frac{\omega}{V}\mu_4 \qquad (3B)$$

$$\frac{d\mu_5}{dt} = 5G\mu_4 - \frac{\omega}{V}\mu_5 \qquad (4B)$$

When a periodic solution is assumed for μ_n, integration of these equations over a cycle gives for Equation (4B)

$$\int_t^{t+\theta} \frac{d\mu_5}{dt}\, dt = \mu_5 \Big|_t^{t+\theta} = 0 = 5\theta\{G\mu_4\} - \frac{\omega}{V}\theta\{\mu_5\} \qquad (5B)$$

$$\{\mu_5\} = 5\frac{V}{\omega}\{G\mu_4\}$$

Substituting this into (2B), we get

$$\{\gamma^2\} = \frac{5\frac{V}{\omega}\mu_3\{G\mu_4\}}{\{\mu_4\}^2} - 1 \qquad (6B)$$

Multiplying (3B) by μ_4 and integrating over the cycle, we get

$$\mu_4 \frac{d\mu_4}{dt} = \frac{1}{2}\frac{d(\mu_4)^2}{dt} = 4G\mu_3\mu_4 - \frac{\omega}{V}\mu_4^2$$

$$\frac{1}{2}\int_t^{t+\theta} \frac{d\mu_4^2}{dt}\, dt = 0 = 4\mu_3\{G\mu_4\} - \frac{\omega}{V}\{\mu_4^2\} \qquad (7B)$$

$$\{G\mu_4\} = \frac{\omega}{V}\{\mu_4^2\}/4\mu_3$$

and substituting (7*B*) into (6*B*)

$$\{\gamma^2\} = 5\,\frac{\{\mu_4^2\}}{\{\mu_4\}^2} - 1 \tag{8B}$$

Subtraction of the steady state γ^2 from $\{\gamma^2\}$ results in

$$\{\gamma^2\} - \gamma^2 = \frac{5}{4}\left[\frac{\{\mu_4^2\}}{\{\mu_4\}^2} - 1\right]$$

The term in parenthesis is the coefficient of variation of μ_4 and since $\{\mu_4^2\} \geq \{\mu_4\}^2$, it is apparent that the accumulated product from a limit cycle will not be tighter than the product from steady state operation. Generally μ_4 does not fluctuate very much and therefore the cycle average $\{\gamma\}$ will be close to the steady state value of 0.5 as shown in Figure 6.

Manuscript received August 18, 1966; revision received January 27, 1967; paper accepted March 28, 1967. Presented at AIChE Philadelphia meeting.

DYNAMIC BEHAVIOR OF THE ISOTHERMAL WELL-STIRRED CRYSTALLIZER WITH CLASSIFIED OUTLET

Martin Sherwin
Reuel Shinnar
and
Stanley Katz

The effect of classification on the dynamic behavior of a crystallizer is investigated by analyzing a simple model of a mixed crystallizer with classified product removal.

It is shown that classification enhances the sensitivity of the crystallizer to disturbances and increases the tendency toward cyclic fluctuations. For a classified product removal model, these fluctuations are observed in the production rate and the magma density.

Industrial crystallization, though one of the oldest chemical engineering unit operations, still remains one of the last to yield to quantitative analysis and is one of the most difficult to operate continuously in a steady state manner. One problem which makes the control of crystallizers difficult is their sensitivity to small upsets and their tendency toward unstable behavior characterized by low frequency cyclic fluctuations of the particle size (*6*, *8*, *11*).

In a previous paper, the possible causes of this tendency toward instability were discussed by analyzing the use of ideally mixed vessel. Though an ideally mixed crystallizer is a strong simplification of a crystallizer, it predicts the correct general behavior of such a system.

Now many crystallizers are operated with some kind of a classification device, either in the crystallizer bed itself or in the outlet of the crystallizer (*6*, *10*). It has been observed that classification increases the tendency toward instability (*6*). With the help of a simple model of an ideally classified crystallizer, this paper attempts to investigate the effect of classification on the dynamic behavior of crystallizers. The steady state behavior of such a system is discussed in a separate paper (*3*) as well as in references *1* and *10*.

The City College of the City University of New York, New York, New York. Martin Sherwin is with Chem Systems, Inc., New York, New York.

In reference *3*, the steady state behavior of this model was observed for an arbitrary recycle ratio S and for a real nonideal classifying device. In the same paper it was shown that in many uses, the assumption made in references *1* and *10*, namely that S is infinite and the classification is ideal, gives a reasonably good approximation of the steady state magma density and product characteristics.

If the particle size separated by the classifier is r_1, then the size distribution of the product in such an ideal classified crystallizer is simply a Dirac delta function at r_1. If it is further assumed that the growth rate is independent of r, then the number density distribution of the crystals in the crystallizer itself is given by a step function (see Figure 1*a*). As the predictions of the steady state behavior of the idealized model are in quite good agreement with those based on solving the more complex case with finite reflux and a nonideal classifier, it was thought that qualitatively at least this ideal model should predict the effect of classification on crystallizer operation fairly well. The following dynamic analysis of a classified crystallizer is therefore based on this simplified model.

DERIVATION OF THE MODEL

A diagram of the type of operation to be discussed is shown in Figure 1*a*. A feed material containing solute at concentration c_0 is fed to the crystallizer at a volu-

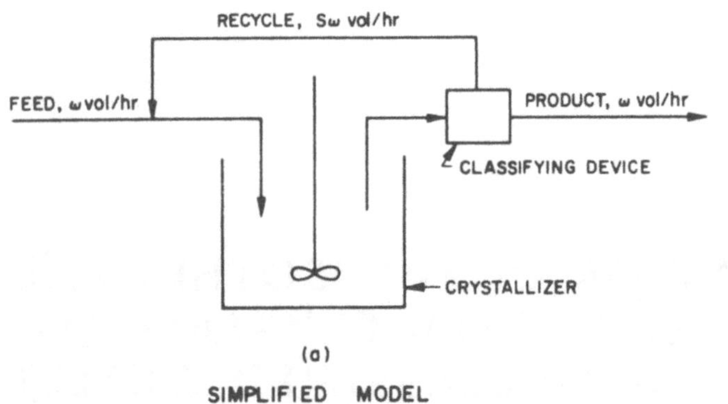

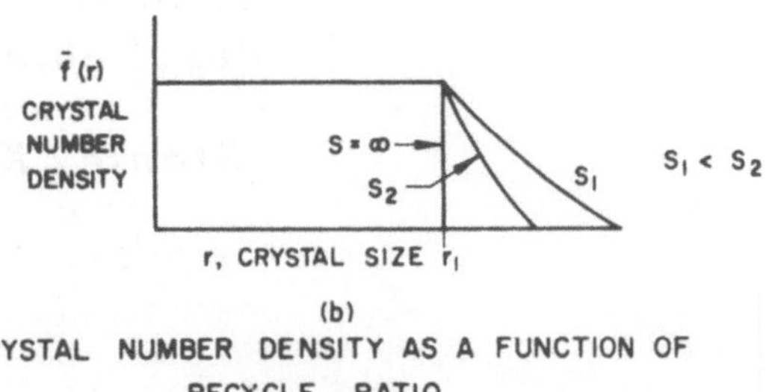

Fig. 1. Classified product crystallizer.

metric rate ω. A slurry is withdrawn from the crystallizer body at a volumetric rate $(S+1)\,\omega$ and sent to a classifying device, the design of which is usually based on centrifugal or elutriative principles. This classifier is designed to produce two separate streams, a product stream of volumetric rate ω which contains all the crystals in the classifier feed which were above a specified critical size r_1, and a recycle stream $S\omega$ which contains all the crystals in the classifier feed stream which were below size r_1.

These classifiers are not so efficient that they perfectly separate all material above and below a specific size r_1. However, the size range of material which is usually dispersed between the classifier product streams is narrow enough to allow this simplification of our model.

In order to describe the behavior of the isothermal mixed suspension classified product crystallizer, balances must be written for the particles and the crystallizing material.

The crystal particles are assumed to be regular solids which maintain their shape upon growth and can therefore be characterized geometrically by a linear dimension r.

For solids which grow in three dimensions, the volume of any crystal is given by kr^3, where k is the appropriate geometrical shape factor. (This paper will deal with the case of a three-dimensional crystal, since it is the most common, although the assumption of plate or needle shaped crystals, which grow in two and one dimensions, respectively, could just as easily be substituted.)

The crystal density ρ is taken to be constant.

The particle size distribution is the number density f, at size r, such that $f(r_1 t)\,dr$ = the number of crystals per unit volume of crystallizer having radii in the range r, $r + dr$ at time t.

The fractional volume ε occupied by the solution is a function of time and can be defined in terms of f:

$$1 - \varepsilon(t) = \int_0^\infty kr^3 f(r, t)\, dr \tag{1}$$

The growth rate of an existing crystal dr/dt is assumed to be described by the product of a function of crystal environment $G(c)$ and a function of crystal size $\phi(r)$ such that

$$\frac{dr}{dt} = G(c)\,\phi(r) \tag{2}$$

The nucleation rate for new crystals is mainly a function of solute concentration in the crystallizer according to the classical theories of Volmer (*16*) and Mier (*17*). It is assumed that the size distribution of new crystals appears as a Dirac delta function, or that all new crystals are formed at a nominal size r_0.

In addition, to simplify the development but not necessarily detract from the model, we also make the following assumptions:

1. $\phi(r) = 1.0$, so that the kinetic growth term given in Equation (2) can be written as

$$\frac{dr}{dt} = G(c)$$

and is only a function of supersaturation.

2. The volumetric feed and product rates are identical, and the volume change accompanying phase change is negligible so the crystallizer working volume V is constant.

3. The feed is a clear solution.

4. The nuclei size r_0 is so small it can be taken as equal to zero. (This is not necessary and was not done during the original work. However, now we state that for the classified product case, the nuclei size has little to no effect on the stability or dynamic behavior of the system. In addition, the assumption $r_0 = 0$ greatly simplifies the following development.)

CONSERVATION EQUATIONS

The appropriate balances can now be written down. (A detailed derivation of the particle balance can be found in reference 5.)

Particle Balance

$$\frac{\partial f(r,t)}{\partial t} + G(c)\,\frac{\partial f(r,t)}{\partial r} = 0 \tag{3}$$

Boundary Conditions

$$f(0, t) = \varepsilon B(c)/G(c)$$

$$f(r, t) = 0; \; r > r_1$$

These terms represent, in order, the accumulation of crystals at size r and the net flux of crystals at size r due to growth. The boundary condition at size zero equates the generation of particles $\varepsilon B(c)$ to those growing away from size zero, as indicated by the term $f(0, t)\, G(c)$. The condition for $r > r_1$ corresponds to the assumption of an infinite recycle ratio. Mathematically speaking, the last is not so much a boundary condition as an expression of our desire to find distributions f which are concentrated in the range $0 \leq r \leq r_1$.

Solute and Crystal Balance

$$\frac{d}{dt}\{V[\varepsilon c + (1 - \varepsilon)\rho]\} = \omega c_0 - \{[\omega - kr_1^3\, Vf(r_1, t)\, G]c - [kr_1^3\, Vf(r_1, t)G]\rho\} \quad (4)$$

These terms represent, in order, the accumulation of solute and crystal, the input of solute by the feed, and the withdrawal of solute and crystal by the classifier product stream. The term $[kr_1^3\, Vf(r_1, t)\, G(c)]$ represents the volumetric rate of crystal product reaching size r_1.

We desire a specific differential equation for the concentration c, since the kinetic terms for nucleation and growth depend on this variable. By multiplying Equation (3) by kr^3 and integrating over r, we get the following relation:

$$\frac{d\varepsilon}{dt} = -\sigma G + kGr_1^3 f(r_1, t) \quad (5)$$

where

$$\varepsilon = 1 - \int_0^\infty kr^3 f\, dr \quad (6)$$

$$\sigma = 3k \int_0^\infty r^2 f\, dr \quad (7)$$

Substituting Equation (5) into (4), we get the following relation for c:

$$\varepsilon \frac{dc}{dt} = \frac{\omega}{V}(c_0 - c) - (\rho - c)\sigma G \quad (8)$$

The working set of equations are now summarized below:

$$\frac{\partial f(r, t)}{\partial t} + G(c)\frac{\partial f(r, t)}{\partial r} = \varepsilon B(c)\, \delta(r) - Gf(r, t)\, \delta(r - r_1)$$

$$\varepsilon \frac{dc}{dt} = \frac{\omega}{V}(c_0 - c) - (\rho - c)\sigma G \quad (9)$$

where $\varepsilon = 1 - \int_0^\infty kr^3 f\, dr$ and $\sigma = 3k \int_0^\infty r^2 f dr$.

MOMENT EQUATIONS

Many times, if it is not necessary to determine how the distribution f varies with time, but a knowledge of the variation of its moments with time is satisfactory, the set can be transformed to a group of ordinary differential equations. In this case, however, the term $f(r_1, t)$, the number density of particles at the cutoff size r_1, prevents the formation of a closed set of moment equations.

Still, the moment equations provide a simple approach to deriving the system's transfer function and subsequently its stability limits.

The nth moment of f will be called μ_n, while the expressions in pointed brackets are averages over f in the same sense; in general

$$<\eta> = \int_0^\infty \eta f\, dr \quad (10)$$

$$\mu_n(t) = \int_0^\infty r^n f(r, t)\, dt \quad (11)$$

The physical significance of the moments of f is shown below:

$\mu_0 = \int_0^\infty f\, dr$ = total number of crystals per unit volume of crystallizer

$\mu_1 = \int_0^\infty rf\, dr$ = total 'radius' of crystals per unit volume of crystallizer

$3k\mu_2 = 3k \int_0^\infty r^2 f\, dr$ = total surface of crystals per unit volume of crystallizer

$k\mu_3 = k \int_0^\infty r^3 f\, dr = (1 - \varepsilon)$ = volume fraction of solids

It should be noted that although the upper limit of integration of r is written as ∞, the term $f(r, t)$ is zero above size r_1.

By multiplying Equation (3) by r^n and by integrating over the range of r there results in the general moments equation

$$\frac{d\mu_n}{dt} = n\mu_{n-1}\, G + \varepsilon\, B\delta(n) - Gf(r_1, t)\, r_1^{\,n}$$

$$n = 0, 1, 2, 3 \ldots \quad (12)$$

In terms of the moments of $f(r, t)$, Equation (8) can be rewritten as

$$\varepsilon \frac{dc}{dt} = \frac{\omega}{V}(c_0 - c) - (\rho - c)\, 3k\mu_2 \quad (13)$$

where

$$\varepsilon = 1 - k\mu_3$$

The leading four moments of (12) plus (13) do not form a closed set because of the term $f(r_1, t)$.

STEADY STATE EQUATIONS

Since we will normalize both the system and linearized equations about their steady states, a diversion to derive the steady state moments is appropriate at this time.

As previously described, the steady state crystal number density is a constant between the nuclei size and the withdrawal size r_1. The crystal number density can be written (superscript bar refers to steady state values) as

$$\bar{f}(r) = \frac{\bar{\varepsilon}\bar{B}}{\bar{G}} \quad 0 \leq r \leq r_1$$
$$\bar{f}(r) = 0 \quad r > r_1 \tag{14}$$

The steady state moments can be directly calculated from (14):

$$\bar{\mu}_n = \int_0^{r_1} r^n \bar{f}(r)\, dr = \frac{\bar{\varepsilon}\bar{B}}{\bar{G}} \frac{r_1^{n+1}}{n+1} \quad n = 0,1,2,3\ldots \tag{15}$$

Substituting (14) into the steady state form of (12), we get a recursion formula for the moments:

$$\bar{\mu}_n = \frac{n}{n+1} r_1 \bar{\mu}_{n-1} \quad n = 1,2,3\ldots \tag{16}$$

It is relatively simple to derive a relation between the product size and solids' residence time τ_s:

$$\tau_s = \frac{\text{Crystal Holdup Volume}}{\text{Crystal Volumetric Production Rate}} \tag{17}$$

$$\tau_s = \frac{k\bar{\mu}_3}{k r_1^3 \bar{G} \bar{f}(r_1)} = \frac{r_1}{4\bar{G}}$$

and

$$r_1 = 4\tau_s G \tag{18}$$

which was derived by Saeman (*10*).

We must distinguish between feed residence time and solids' residence time and voidage fraction in the crystallizer and voidage fraction in the product, which usually are not the same for the classified crystallizer.

We denote the voidage fraction of the product stream by $\bar{\varepsilon}_p$, and from a solute and crystal balance around the system one obtains

$$\bar{\varepsilon}_p = \frac{\rho - c_0}{\rho - \bar{c}} \tag{19}$$

and

$$(1 - \bar{\varepsilon}_p) = \frac{k r_1^3 \bar{\varepsilon}\bar{B} V}{\omega} \tag{20}$$

where $k r_1^3 \bar{\varepsilon}\bar{B} V$ is equal to the volumetric rate of production of crystals of size r_1.

From (17) and (18) we can also write

$$\tau_s = \frac{(1-\varepsilon)V}{(1-\varepsilon_p)\omega} = \frac{(1-\varepsilon)}{(1-\varepsilon_p)}\tau \tag{21}$$

from which we see that when the crystallizer body voidage and product voidage are equal, the solids and feed residence times are also equal. This is always true for the mixed crystallizer with no classification.

LINEARIZATION OF THE MODEL

We first studied the linearized system to determine the regions of stable and unstable behavior.

The feed concentration can vary in the form

$$c_0(t) = \bar{c}_0 + c_0'(t)$$

The performance variables are represented by

$$\mu_n(t) = \bar{\mu}_n + \mu_n'(t) \quad n = 0,1,2,3\ldots$$
$$c(t) = \bar{c} + c'(t)$$
$$f(r,t) = \bar{f}(r,t) + f'(r,t)$$

The initial conditions are

$$c(0) = \bar{c}$$
$$\mu_n(0) = \bar{\mu}_n \quad n = 0,1,2,3\ldots$$
$$f(r,0) = \bar{f}(r) \tag{22}$$

The kinetic terms for small deviations from steady state can be written as

$$B(c) = \bar{B} + \frac{dB}{dc}(\bar{c})\, c'$$

$$G(c) = \bar{G} + \frac{dG}{dc}(\bar{c})\, c'$$

The following dimensionless variables are defined:

$$Z_n'(\theta) = \frac{\mu_n'}{\bar{\mu}_n} \quad n = 0,1,2,3\ldots$$

$$Y'(\theta) = \frac{\varepsilon c'}{\bar{c}_0 - \bar{c}}$$

$$\mathcal{F}'(r,\theta) = \frac{f'(r,t)}{\bar{f}}$$

$$\theta = t/\tau_s$$

$$q(\theta) = \frac{c_0'}{\bar{c}_0 - \bar{c}}$$

$$b = \frac{\tau}{\tau_s} \frac{\bar{c}_0 - \bar{c}}{\bar{\varepsilon}\bar{B}} \frac{dB}{dc}(\bar{c})$$

$$g = \frac{\tau}{\tau_s} \frac{\bar{c}_0 - \bar{c}}{\bar{\varepsilon}\bar{B}} \frac{dB}{dc}(\bar{c})$$

$$\frac{b}{g} = \frac{\frac{dB}{dc}(\bar{c})}{\frac{\bar{G}}{\bar{B}}\frac{dG}{dc}(\bar{c})} \tag{23}$$

Substituting relations (22) and (23) into (12) and (13), we get the following linearized equations:

$$\frac{dZ_n'}{d\theta} - \left[\left(n\frac{\bar{\mu}_{n-1}}{\bar{\mu}_n}\tau_s\bar{G}\right)g + \left(\frac{\bar{\varepsilon}\bar{B}\tau_s}{\bar{\mu}_n}\delta(n)\right)b - \left(\frac{\bar{\varepsilon}\bar{B}\tau_s r_1^n}{\bar{\mu}_n}\right)g\right]Y' - \left(n\frac{\bar{\mu}_{n-1}}{\bar{\mu}_n}\tau_s\bar{G}\right)Z_{n-1}' + \left(kB\frac{\bar{\mu}_3}{\bar{\mu}_n}\tau_s\delta(n)\right)Z_3' + \left(\frac{\bar{\varepsilon}\bar{B}\tau_s r_1^n}{\bar{\mu}_n}\right)\mathcal{F}'(r_1,\theta) = 0$$

$$n = 0,1,2,3\ldots \tag{24}$$

$$\frac{dY'}{d\theta} + \left(\frac{3\bar{\mu}_2\bar{G}}{\bar{\varepsilon}\bar{B}r_1^3}\right)g + \frac{1}{\bar{\varepsilon}}\left(\frac{\rho - \bar{c}_0}{\rho - \bar{c}}\right)Y' + \left(\frac{3\bar{G}\bar{\mu}_2}{\bar{\varepsilon}\bar{B}r_1^3}\right)Z_2' = q(\theta) \tag{25}$$

The groups contained within parenthesis in Equations (24) and (25) are composed solely of steady state values and can easily be calculated from Equations (8) and (14) to (18). By performing this exercise, the linearized system equations reduce to the following:

$$\frac{dZ_0'}{d\theta} - \frac{(b-g)}{4}Y' + \frac{(1-\bar{\varepsilon})}{4\bar{\varepsilon}}Z_3' - \frac{\mathcal{F}'}{4}(r_1,\theta) = 0$$

$$\frac{dZ_1'}{d\theta} - \frac{1}{2}Z_0' + \frac{1}{2}\mathcal{F}'(r_1,\theta) = 0$$

$$\frac{dZ_2'}{d\theta} - \frac{3}{4}Z_1' + \frac{3}{4}\mathcal{F}'(r_1,\theta) = 0$$

$$\frac{dZ_3'}{d\theta} - Z_2' + \mathcal{F}'(r_1,\theta) = 0$$

$$\frac{dY'}{d\theta} + \left[\frac{\tau_s/\tau - (1-\bar{\varepsilon})}{\bar{\varepsilon}} + g\right]Y' + Z_2' = q(\theta) \tag{26}$$

CONCENTRATION DEPENDANCE OF CRYSTAL GROWTH AND NUCLEATION

Equation (26) contains the dimensionless groups b and g. The value of these groups at steady state depends on the sensitivity of the nucleation rate and crystal growth rate to a change in supersaturation. Now Equation (26) holds for any case in which our original simplifying assumptions hold (complete mixing, perfect classification, B and G are functions of C only) and are independent as to the exact nature of the dependance of B and G (or b and g) on supersaturation. To be exact, one could quite easily devise a set of equations on which G depends on r and B depends on the moments of the size distributions of crystals in the vessel. The equations would, however, be much harder to solve, and it was therefore thought worthwhile to investigate the simpler case for which B and G depend on concentration only.

It should be mentioned here that the later sections will show that the ratio of the two groups b/g is of primary importance in determining the stability and dynamic behavior of the model. b/g represents the sensitivity ratio of the nucleation to growth functions (at steady state), and the form depends on whether nuclei are removed by a fines trap, added by seed addition, or just generated internally.

Now, while Equation (26) can easily be treated for any dependence of G on c, many investigators (*6, 10, 14*) have found the relation between G and c to be often satisfied by

$$G(c) = k_1(c - c_s) \tag{27}$$

Substituting this model into Equation (23), we get

$$g = \frac{\tau}{\tau_s\bar{\varepsilon}}\frac{\bar{c}_0 - \bar{c}}{\bar{c} - c_s} \tag{28}$$

Generally, the supersaturation $(c - c_s)$ of crystallizing systems is quite small when compared with the difference between feed and outlet concentrations, so that normally the value of g should be of the order of magnitude 10^2 to 10^3. The dependance of B on supersaturation is more complex and has been discussed in our previous papers (*3, 11*). Nucleation inside a stirred crystallizer is a complex phenomenon in which crystallization occurs by attrition, homogeneous crystallization in the solution, and heterogeneous nucleation at the crystal surface. We neglect here the case in which nucleation is mainly due to attrition. Now, both heterogeneous and homogeneous nucleation depend on the supersaturation, and at very low supersaturation this dependance is highly nonlinear. The authors, therefore, chose for the purposes of this investigation to use as the nucleation rate Volmer's expression

$$B(c) = k_2\, e^{-\frac{k_3}{[\ln c/c_s]^2}} \tag{29}$$

Volmer's original equation was derived for homogeneous nucleation, which does not necessarily apply here. In our case, it is used mainly as a nonlinear smooth fitting function which expresses the nonlinear form of $B(c)$ at very low supersaturations. This is similar to the way that an Arrhenius relation expresses the temperature dependance of the reaction rate for complex reactions, even though the real reactions are far more complex than a simple rate equation indicates. In heterogeneous reactions, the reaction rate also depends on the amount of foreign nuclei and crystal surface, but in a crystal magma their amount is so great that the dependance of B on them is slight as compared with the effect of slight changes in c.

Substituting Equation (29) into Equation (23), we obtain

$$b = \frac{\tau}{\tau_s \bar{\varepsilon}} \frac{2 k_3 (\bar{c}_0 - \bar{c})}{\bar{c}[\ln c/c_s]^2} \approx \frac{\tau}{\tau_s \bar{\varepsilon}} \frac{2 k_3}{\bar{c}\left(\frac{\bar{c}}{c_s} - 1\right)^3} \tag{30}$$

the latter approximation being made since $[(\bar{c}/c_s - 1)]$ is very small for the type of crystallization system under consideration.

Then, by using Equations (27) and (28), the term b/g can be approximated by

$$\frac{b}{g} \approx \frac{2 k_3 (\bar{c} - c_s)}{\left(\frac{c}{c_s} - 1\right)^3 \bar{c}} = \frac{2 k_3 \, c_s}{\left(\frac{\bar{c}}{c_s} - 1\right)^2 \bar{c}} \approx \frac{2 k_3}{\left(\frac{c}{c_s} - 1\right)^2} \tag{31}$$

The term b/g given in Equation (31) represents the sensitivity ratio of the nucleation to growth functions at steady state conditions without any external source of nuclei addition or removal, and it can be seen that as the solute concentration approaches its solubility limit, this term rapidly increases in value. This means that as one tries to grow larger crystals, a critical size should be reached where the stability limits are exceeded.

Another nucleation model which is often used to fit experimental data is Mier's metastable moded (*17*), which can be written as

$$B = k_4 (c - c_M)^m \qquad c > c_M$$
$$B = 0 \qquad c \leq c_M \tag{32a}$$

where k_4 is a constant, c_M is the metastable concentration, above which nucleation occurs, and m is some power. Combining this with Equation (27) we get

$$\frac{b}{g} = m \frac{\bar{c} - c_s}{\bar{c} - c_M} \qquad c > c_M \tag{32b}$$

Here it is evident that as the solute concentration approaches the metastable limit c_M from above, the sensitivity ratio b/g will increase in value. A more detailed discussion of the applicability of these nucleation expressions to a real crystallizer is given in the previous paper (*11*). It should also be pointed out that the stability analysis itself is independent as to the form of the assumed dependance of B on c.

Many crystallizers with classified outlets have some device to reduce the actual nucleation rate by a nuclei trap (*10*). The effect of a nuclei trap on $B(c)$ is discussed in references(*3* and *11*). One can either treat the nuclei trap by changing the constant in Equation (2), or one might more accurately take into account the effect of G on the efficiency of the nuclei trap. This can be done as follows. A fines trap usually consists of withdrawing solution from some part of the vessel in such a manner that only crystals below a certain size will be elutriated with the solution. This stream is heated to redissolve the solids and then returned to the crystallizer body. Most often this operation is carried on continuously rather than on the basis of some feedback signal.

In order to dissolve a nuclei, it must be trapped very early, before it has a chance to grow, say within ⅛ or ¼ or a drawdown time τ. Over this time period, even for cycling or unsteady behavior, the variation in supersaturation or growth rate $G(c)$ is relatively small. We therefore assume a quasi steady state conditions such that the fraction of new nuclei being generated at a specific time which survive the trap can be related to the instantaneous growth rate.

If τ' is the ratio of the total crystallizer volume to the flow rate through the fines trap, then the probability of a growing crystal to survive to such a size larger than the critical size of the fines trap r_c is

$$\text{fraction of nuclei growing to } r_c = e^{-r_c/\bar{G}\tau'}$$

since we have made the quasi steady state assumption over this small time period (*3*). Therefore, the net nucleation rate for the crystallizer and associated nuclei trap is

$$\text{net nucleation} = B(c)\, e^{-r_c/\bar{G}\tau'} \tag{33a}$$

$$b/g = \frac{\bar{G} \frac{dB}{dc}(\bar{c})}{\bar{B} \frac{dG}{dc}(\bar{c})} + \frac{r_c}{\tau' \bar{G}} \tag{33b}$$

whereas for operation without the fines trap or other auxilliaries

$$b/g = \frac{\bar{G} \frac{dB}{dc}(\bar{c})}{\bar{B} \frac{dG}{dc}(\bar{c})} \tag{23}$$

all the terms being evaluated at the steady state. Other ways of operating the fines trap are treated in similar ways.

It should be noted that in this analysis the total nucleation rate is based on the free volume. Though this is consistent with assuming that the nucleation rate depends on (c), it is not correct for those cases in which the nucleation rate depends on total crystal surface. A similar treatment can be applied to the more complex case, where B depends on the moments of the particle size distribution inside the crystallizer.

STABILITY ANALYSIS

The set of Equations (26) cannot yet be analyzed for stability, since an additional expression for $\mathfrak{F}(r_1, \theta)$ is required. To develop this expression, we return to the particle balance Equation (3). Linearizing and normalizing this expression with the aid of relations (22) and (23), we get

$$\frac{\partial \mathfrak{F}'(r, \theta)}{\tau_s \overline{G} \partial \theta} + \frac{\partial \mathfrak{F}'(r, \theta)}{\partial r} = 0 \qquad (34)$$

$$\text{B.C. } \mathfrak{F}'(0, \theta) = (b - g) Y' - \left(\frac{1 - \overline{\varepsilon}}{\overline{\varepsilon}}\right) Z_3'$$

It is obvious that we cannot get an ordinary differential expression for $\mathfrak{F}(r_1, \theta)$ from (34). Therefore, we turn to the transfer function to examine the system's stability boundaries. The partial differential equation indicates that the transfer function will contain an exponential function.

Taking the Laplace transform of Equation (34) with respect to θ and denoting the transform by the superscript $\hat{}$, we get

$$\frac{s\hat{\mathfrak{F}}'(r, s)}{\tau_s G} + \frac{\partial \hat{\mathfrak{F}}'(r, s)}{\partial r} = 0 \qquad (35)$$

$$\text{B.C. } \hat{\mathfrak{F}}'(0, s) = (b - g) \hat{Y}(s') - \left(\frac{1 - \overline{\varepsilon}}{\overline{\varepsilon}}\right) \hat{Z}_3'(s)$$

Solving (35), we get

$$\hat{\mathfrak{F}}'(r, s) = (b - g) \hat{Y}(s') - \left(\frac{1 - \overline{\varepsilon}}{\overline{\varepsilon}}\right) \hat{Z}_3'(s) \; e^{-\frac{sr}{\tau_s \overline{G}}} \qquad (36)$$

and using Equation (18) we finally get an expression for $\hat{\mathfrak{F}}'(r_1, s)$:

$$\hat{\mathfrak{F}}'(r_1, s) = (b - g) \hat{Y}'(s) - \left(\frac{1 - \overline{\varepsilon}}{\overline{\varepsilon}}\right) \hat{Z}_3'(s) \; e^{-4s} \qquad (37)$$

Combining (37) with the Laplace transforms of (26), we get the following characteristic matrix:

Multiplying out the corresponding determinant, we find that the denominator of any transfer function will have the following characteristic equation:

$$A_5 S^5 + A_4 S^4 + A_3 S^3 + A_2 S^2 + A_1 S + A_0 + e^{-4S} (B_4 S^4 + B_3 S^3 + B_2 S^2 + B_1 S + B_0) = 0 \qquad (39)$$

The coefficients of this term are functions of the steady state parameters $b, g, \overline{\varepsilon}$, and τ_s/τ.

The stability limits of this transfer function were investigated by use of Nyquist diagrams, actual solution for the roots, and the determination of crossover points of the roots over the imaginary axis. These techniques are described in detail in reference *2* and are further discussed in Appendix 1.

As for the unclassified case (*11*), it was found that the term b/g was the crucial parameter which determined the stability of the system. It was found that the stability contour of b/g was insensitive to changes in g (for value greater than 10^2) and τ_s/τ but did vary with changes in $\overline{\varepsilon}$. Below values of 0.6, ε has a rather strong effect, but such low values of ε are of no practical interest. These limits are shown in Figure 2. It is seen that the critical values of b/g are about 1 to 3 for classified product operation, whereas for the unclassified operation, they were about equal to 20 (*11*). The addition of a classifying device greatly decreases the systems stable operating range.

EFFECT OF OPERATING VARIABLES ON STABILITY

The independent variables that an operator has at his control for our model classified crystallizer are temperature, feed rate, product size, and fraction of nuclei destroyed (if a fines trap is included). The present paper treats an isothermal case so that the temperature effect has been disregarded. (It should be pointed out that although some papers deal with the effect of tem-

Characteristic Matrix

Variable → / Equation ↓	$\hat{Z}_0'$	$\hat{Z}_1'$	$\hat{Z}_2'$	$\hat{Z}_3'$	$\hat{Y}'$	$\hat{\mathfrak{F}}'(r_1, s)$	Disturbances
0 moment	S	0	0	$\frac{1-\overline{\varepsilon}}{\varepsilon}\left(\frac{1}{4}\right)$	$-(b-g)\frac{1}{4}$	$\frac{1}{4}$	0
1 moment	$-\frac{1}{2}$	S	0	0	0	$\frac{1}{2}$	0
2 moment	0	$-\frac{3}{4}$	S	0	0	$\frac{3}{4}$	0
3 moment	0	0	-1	S	0	1	0
Solute Bal.	0	0	1	0	$s + g + \frac{[\tau_s/\tau - (1-\overline{\varepsilon})]}{\overline{\varepsilon}}$	0	$\hat{q}(s)$
$\hat{\mathfrak{F}}'(r_1, s)$	0	0	0	$\frac{1-\overline{\varepsilon}}{\overline{\varepsilon}}$	$-(b-g)$	e^{4s}	0

(38)

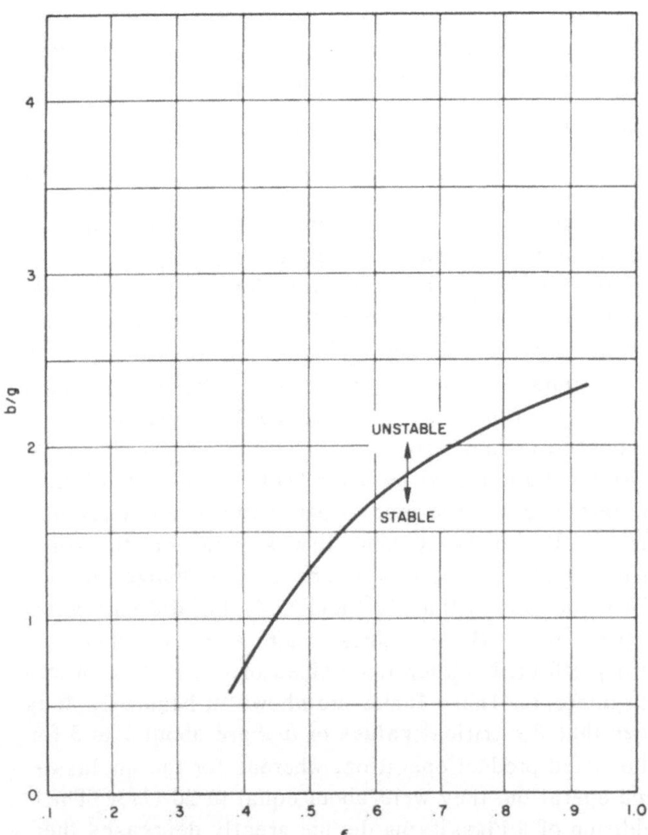

Fig. 2. Stability limits for the well-mixed classified product crystallizer.

perature on growth rate, there is very little similar information for nucleation of solute from solution.)

From the operators point of view it is important to consider the effect of varying production rate with constant product size and the effect of varying product size with constant production rate on the system's stability.

Varying Production Rate

Since the supersaturation is so small when compared with the concentration change between feed and product, we can assume that the volume fraction of solids in the product stream will remain unchanged. From Equation (20) we see that by assuming constant product size r_1, this means (subscripts 1 and 2 refer here to different operation conditions)

$$\frac{\overline{\varepsilon}_1\overline{B}_1}{\overline{\omega}_1} = \frac{\overline{\varepsilon}_2\overline{B}_2}{\overline{\omega}_2} \tag{40}$$

Also, constant product size denotes [from Equations (18) and (21)]

$$\tau_{s_1}\overline{G}_1 = \tau_{s_2}\overline{G}_2$$

$$(1-\overline{\varepsilon}_1)\frac{\overline{G}_1}{\overline{\omega}_1} = (1-\overline{\varepsilon}_2)\frac{\overline{G}_2}{\overline{\omega}_2} \tag{41}$$

Solving for $\overline{\omega}_1/\overline{\omega}_2$ from (41) and substituting into (40), we get

$$\frac{\overline{\varepsilon}_1\overline{B}_1}{(1-\overline{\varepsilon})\overline{G}_1} = \frac{\overline{\varepsilon}_2\overline{B}_2}{(1-\overline{\varepsilon}_2)\overline{G}_2} = \text{constant} \tag{42}$$

Assuming a linear growth model as given in Equation (27) and a Volmer type of nucleation model as given in Equation (29), we can use the definition given for b/g in Equation (31). In terms of b/g, the ratio for growth and nucleation rates takes the following forms:

$$\frac{\overline{G}_1}{\overline{G}_2} = \left[\frac{(b/g)_2}{(b/g)_1}\right]^{1/2} \tag{43}$$

$$\frac{\overline{B}_1}{\overline{B}_2} = e^{1/2[(b/g)_2-(b/g)_1]} \tag{44}$$

Substituting (43) into (44) into (42), we obtain

$$\frac{\overline{\varepsilon}_1}{1-\overline{\varepsilon}_1}(b/g)_1^{1/2}\,e^{-1/2(b/g)_1} =$$

$$\frac{\overline{\varepsilon}_2}{1-\overline{\varepsilon}_2}(b/g)_2^{1/2}\,e^{-1/2(b/g)_2} = \text{constant} \tag{45}$$

Starting with a stable base point, say $\varepsilon_1 = 0.70$, $(b/g)_1 = 1.50$, we can solve (45) for other combinations of ε_2 and $(b/g)_2$ and then from either (40) or (41) determine the relative production rate $\overline{\omega}_2/\overline{\omega}_1$.

Figure 3 plots this curve of b/g vs. ε for varying production rates through the base point $\varepsilon_1 = 0.70$, $(b/g)_1 = 1.50$. The stability boundary is shown on the same graph. It can be seen that if the original system is stable; decreasing the production rate while maintaining a constant product size will lead to an unstable situation.

It should, however, be remembered that production rate can only be changed over a relatively narrow range. It should also be noted that while increasing the production rate (when possible) increases c and therefore decreases b/g, it also sometimes reduces the quality of the product (*10*).

Varying Product Size

By the same reasoning as before, we expect the solids content of the product stream to be essentially constant. And from Equation (20) with constant production rate, we see

$$(r_1)_1^3\,\overline{\varepsilon}_1\overline{B}_1 = (r_1)_2^3\,\overline{\varepsilon}_2\overline{B}_2 = \text{constant} \tag{46}$$

From Equation (18) it can also be shown that

$$(r_1)_1/(1-\overline{\varepsilon}_1)\overline{G}_1 = (r_1)_2/(1-\overline{\varepsilon}_2)\overline{G}_2 = \text{constant} \tag{47}$$

Using Equations (43) and (44) we can follow the same method to determine the variation of ε and b/g and relative size, when starting with the same base point. This is also shown in Figure 2.

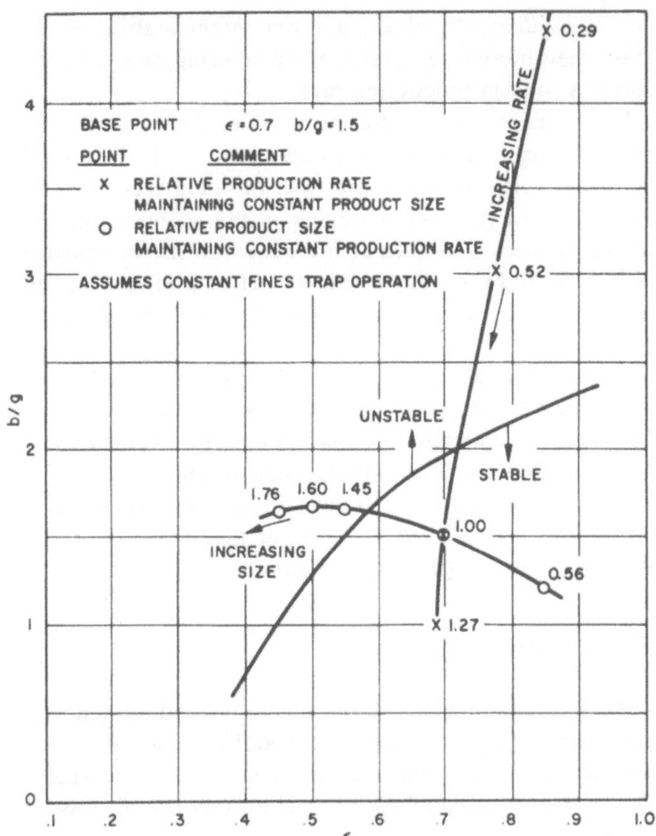

Fig. 3. Variation of b/g and ε with changing product size or production rate.

It can be seen that increasing the product size while maintaining a constant production rate (and constant fines trap operation) will lead to an unstable situation.

If it is desired to maintain a constant growth rate (constant b/g) when varying production rate or product size, a change in fines trap operation would be necessary to remove or redissolve the excess nuclei formed. A similar analysis shows that a smaller fraction of nuclei should be dissolved in the fines trap when increasing production (constant product size), and a larger fraction of nuclei should be dissolved in the fines trap when increasing the size of the crystals in the product while maintaining constant production rate.

COMPARISON OF MIXED VS. CLASSIFIED OPERATION

We now try to compare the relative stability of a mixed outlet vs. a classified outlet crystallizer. This is somewhat difficult, as the well-stirred crystallizer with a mixed outlet for given c_0 and c_s is a uniquely defined system, whereas in the case of a classified crystallizer there is some flexibility (see reference *3*). Furthermore, a classified crystallizer requires a nuclei trap more often than a crystallizer with a mixed outlet, which again makes the comparison more difficult.

One way of comparing them is at the same production rate. In this case to achieve the same weight average particle size the total net nucleation rate in the classified crystallizer must be 0.094 of the rate in the mixed outlet crystallizer. This reduced nucleation rate can be achieved by the following.

1. Increasing the solids residence time (which is the same as reducing the free volume), thereby increasing μ_3 and reducing the supersaturation.
2. Using a fines trap to reduce the effective nucleation rate at constant supersaturation.
3. A combination of both methods. In alternative 1 and 3 the supersaturation is lower than in the mixed case. If we are operating near the metastable limit, this means that normally b/g will be higher for the classified case than for the mixed case.

The effect of a nuclei trap is more difficult to evaluate directly and will be discussed in a future paper.

As the critical value of b/g for the classified product case is much lower than for the mixed product case, this implies that operating with a mixed outlet almost always reduces the tendency toward instability and slow frequency oscillation.

EFFECT OF SEED ADDITION

Assuming that the seed size is identical to the generated nuclei size (both distributions appearing as delta functions), the results of the preceding sections can be used to predict the effect of seed addition on the system's stability. The nucleation term $B(c)$ is expanded to include all nuclei entering the system as shown below:

$$B(c)_{\text{Total}} = B(c)_{\text{nucleated}} + B_{\text{seed}} \tag{48}$$

If we represent the steady state number of seed as equals to ξ times the generated nuclei (superscript bar indicates steady state values)

$$B_{\text{seed}} = \xi \overline{B}(c)_{\text{nucleated}} \tag{49}$$

so that

$$\overline{B}(c)_{\text{Total}} = (\xi + 1)\,\overline{B}(c)_{\text{nucleated}} \tag{50}$$

the sensitivity parameter of the system b/g for the seeded case will be less than the sensitivity parameter for the unseeded case at identical supersaturations:

$$\frac{b}{g}_{\text{seeded system}} = \frac{1}{1+\xi}\,\frac{b}{g}_{\text{unseeded system}} \tag{51}$$

It was shown that for the mixed case (*11*) seeding increased the stability of the system with little deleterious effect on product size. This was because the seeding increased the crystal surface area which lowered the supersaturation and hence the nucleation rate. Owing to this compensating effect, the resulted change in nucleation rate was slight.

In the classified crystallizer we have control over the crystal surface area (or magma density) and can make the trade of seeded nuclei for generated nuclei more easily than in the mixed case.

To illustrate, let us assume we are operating a classified product crystallizer to produce the same weight mean size crystal as was realized in a mixed product operation. One method for achieving this is for a fines trap to be added to the system. The sensitivity expression becomes

$$(b/g) = \frac{\overline{G}}{\overline{B}} \frac{\frac{dB}{dc}(c)}{\frac{dG}{dc}(c)} + \frac{r_c}{\tau' \overline{G}} \tag{52}$$

And if the system is seeded, the system sensitivity parameter becomes

$$(b/g) = \frac{\frac{\overline{G}}{\overline{B}} \frac{\frac{dB}{dc}(c)}{\frac{dG}{dc}(c)} + \frac{r_c}{\tau' \overline{G}}}{1 + \xi} \tag{53}$$

Say the classified product system is unstable (for example, $\varepsilon = 0.9$, $\frac{G}{B}\frac{dB/dc}{dG/dc}(\overline{c}) = 5$, $r_c/\tau'\overline{G} = 2.36$, and $(b/g) = 7.36$). Without seeding there is not much we could do to stabilize the operation while still producing the same volume and size product (other than install a feedback control system). With seeding, we could raise the solids fraction and reduce the supersaturation so that only a minor portion of the nuclei are generated owing to the supersaturation of the solution.

To produce the same size and number of crystals as was originally intended, the net nuclei introduced both internally and externally must be the same:

$$\varepsilon B = \text{constant } (\omega \text{ is the same}) \tag{54}$$

From (41)

$$\tau_s \overline{G} = \text{constant}$$
$$(1 - \overline{\varepsilon})\overline{G} = \text{constant} \tag{55}$$

By lowering ε to 0.8 we can use Equations (41), (43), and (44) to calculate that now $\frac{G}{B}\frac{dB/dc}{dG/dc}(\overline{c}) = 20$, and the generated nuclei are a very small factor indeed $\approx 5(10^{-3})$ of the original number. The nuclei are now almost exclusively supplied by seed [calculated from (40)], and the fines trap need only be operated if the amount of seed is excessive:

$$\text{if } \xi = 198 \quad \frac{r_c}{\tau'\overline{G}} = 0 \quad \text{and} \quad (b/g) = 0.1$$

$$\text{if } \xi = 500 \quad \frac{r_c}{\tau'\overline{G}} = 0.91 \quad \text{and} \quad (b/g) = 0.04$$

It is evident, therefore, that one might stabilize the classified product crystallizer by seeding and yet not suffer a loss in production rate.

It was mentioned in the introduction and in references *3* and *11* that real crystallizers, even when highly agitated, sometimes deviate from the idealized model, as part of the nucleation occurs in the nonmixed entrance region. This effect increases as the crystallizer is scaled up (*12*). The net effect of this in homogeneity is the same as seeding and should have a stabilizing effect.

A similar effect to that of seeding could be achieved by staging the crystallizer, by adding a small mixed first stage whose particle size is small as compared with r_1 and is therefore stable, and which acts as a stabilizing seed source to the crystallizer with classified outlet. A rigorous analysis of these last cases is, however, outside the scope of this paper.

NONLINEAR SOLUTIONS

We wanted solutions to the nonlinearized system equations in time to check the model's behavior in the regions outlined by the stability analysis and to study the characteristics of the limit cycles in the unstable region.

The starting point of this work was the set of Equations (9) reproduced below:

Particle balance

$$\frac{\partial f(r,t)}{\partial t} + G(c)\frac{\partial f(r,t)}{\partial r} = 0$$

B. C. (56)

$$f(0,t) = \frac{\varepsilon B(c)}{G(c)}, \quad f(r > r_1, t) = 0$$

Concentration balance

$$\varepsilon \frac{dc}{dt} = \frac{\omega}{V}(c_0 - c) - (\rho - c)\,3kG(c)\mu_2 \tag{57}$$

We did not attempt to work with the moment equations for this study, since we saw before that for this case they do not lead to a closed set of ordinary differential equations. Therefore, the solution of a partial differential equation will be required to follow the dynamics of this system.

To simplify the study of the dynamic behavior of the system, the variables were normalized about the steady state variables corresponding to the feed concentration c_0 and rate ω.

The following dimensionless variables and functions are defined:

$$\mathcal{F}(\ell, \theta) = \frac{f(\ell,\theta)}{\tilde{f}} = \frac{f(\ell,\theta)}{\varepsilon \overline{B}/\overline{G}} \tag{58}$$

$$Y = \frac{c - c_s}{\bar{c} - c_s}$$

$$\ell = r/r_1$$

$$\theta = t/\tau_s$$

$$Z_n = \frac{\mu_n}{\bar{\mu}_n} \qquad n = 0, 1, 2, 3 \ldots$$

Substitution of (58) into (57) along with (15) to (18) results in the following dimensionless equations:

Normalized particle balance

$$\frac{\partial \mathcal{F}(\ell,\theta)}{\partial \theta} + \frac{1}{4}\frac{G}{\bar{\bar{G}}}\frac{\partial \mathcal{F}(\ell,\theta)}{\partial \ell} = 0$$

B.C.

$$\mathcal{F}(0,\theta) = \frac{(B/\bar{B})}{(G/\bar{G})}\frac{[1 - (1 - \bar{\varepsilon})Z_3]}{\bar{\varepsilon}} \qquad (59)$$

$$\mathcal{F}(\ell > 1, \theta) = 0$$

Normalized concentration balance

$$(1 - (1 - \bar{\varepsilon})Z_3)\frac{dY}{d\theta} = \frac{\tau_s}{\tau}\left[\frac{c_0 - c_s}{\bar{c} - c_s} - Y\right] - \left[\frac{\rho - c_s}{c - c_s} + 1 - Y\right]\left[\frac{G}{\bar{G}}(1 - \bar{\varepsilon})Z_2\right] \qquad (60)$$

where

$$Z_n = (n+1)\int_0^1 \ell^n \mathcal{F}(\ell,\theta)\,d\ell \qquad n = 0, 1, 2, 3 \ldots \qquad (61)$$

The two concentration terms in Equation (60)

$$\frac{\bar{c}_0 - c_s}{\bar{c} - c_s} \quad \text{and} \quad \frac{\rho - c_s}{\bar{c} - c_s}$$

can be rearranged in terms of groups previously defined in the stability study. By taking the kinetic growth model to be linear in supersaturation

$$G(c) = k_1(c - c_s) \qquad (27)$$

Then, as before

$$g = \frac{\tau_s}{\tau}\frac{1}{\bar{\bar{\varepsilon}}}\frac{\bar{c}_0 - \bar{c}}{\bar{c} - c_s} \qquad (28)$$

and

$$\frac{\bar{c}_0 - c_s}{\bar{c} - c_s} = \frac{\bar{c}_0 - \bar{c}}{\bar{c} - c_s} + 1 = 1 + \bar{\varepsilon}g\frac{\tau}{\tau_s} \qquad (62)$$

By substituting the steady state relation of (57) into the second group

$$\frac{\rho - c_s}{\bar{c} - c_s} = \frac{\rho - \bar{c}}{\bar{c} - c_s} + 1 = \frac{\tau_s}{\tau}\frac{c_0 - \bar{c}}{\bar{c} - c_s} + 1 = \frac{\bar{\varepsilon}}{1 - \bar{\varepsilon}}g + 1 \qquad (63)$$

Also

$$\frac{G}{\bar{\bar{G}}} = \frac{k_1(c - c_s)}{k_1(\bar{c} - c_s)} = Y \qquad (64)$$

By using the Volmer nucleation model and by remembering that we are assuming that c/c_s is close to 1, the nucleation rate can be written as

$$B = k_2 e^{\frac{-k_3}{[\ell_n c/c_s]^2}} \approx k_2 e^{-\frac{k_3}{(c/c_s - 1)^2}} \qquad (65)$$

Then, substituting (31) into this, we get

$$\left(\frac{B}{\bar{\bar{B}}}\right) = e^{\frac{1}{2}\frac{b}{g}\left(1 - \frac{1}{Y^2}\right)} \qquad (66)$$

With the incorporation of these kinetic models and substitution of (62), (63), (64), and (66) into (60), a working set of equations assumes the following form:

Normalized particle balance

$$\frac{\partial \mathcal{F}(\ell,\theta)}{\partial \theta} + \frac{1}{4}Y\frac{\partial \mathcal{F}(\ell,\theta)}{\partial \ell} = 0$$

B.C.

$$\mathcal{F}(0,\theta) = \frac{(B/\bar{B})}{Y}\frac{[1 - (1 - \bar{\varepsilon})Z_3]}{\bar{\varepsilon}}, \quad \mathcal{F}(\ell > 1, \theta) = 0 \qquad (67)$$

Normalized concentration balance

$$[1 - (1 - \bar{\varepsilon})Z_3]\frac{dY}{d\theta} = \left[\bar{\varepsilon}g + (1 - Y)\frac{\tau_s}{\tau}\right] - \left[\frac{\bar{\varepsilon}}{1 - \bar{\varepsilon}}g + 1 - Y\right][Y(1 - \bar{\varepsilon})Z_2] \qquad (67)$$

where

$$Z_n = (n+1)\int_0^1 \ell^n \mathcal{F}(\ell,\theta)\,d\ell \qquad n = 0, 1, 2, 3 \ldots$$

$$\frac{B}{\bar{\bar{B}}} = e^{\frac{1}{2}\frac{b}{g}\left[1 - \frac{1}{Y^2}\right]}$$

The steady state parameters which determine the normalized dynamic behavior of the system are $\bar{\varepsilon}$, τ_s/τ, g, and b/g, the same groups which determined the stability regions in the linearized analysis.

As in the linearized study, it was found that changing the values of τ_s/τ and g had a very small effect on the normalized dynamic behavior, and the parameters $\bar{\varepsilon}$ and b/g were of primary significance.

Numerical solutions corroborated the linearized analysis in that for all systems where parameters indicated stable operation, initial perturbations damped out, regardless of the size of the perturbation. As expected for all systems whose parameters indicated linear instability, initial perturbations from steady state grew into well-defined limit cycles.

The characteristics of a typical limit cycle are shown in Figure 4. Figure 5 illustrates the fluctuations in the normalized particle densities over the limit cycle.

From Figure 4 we see that the same internal feedback mechanism is prevalent in the classified operation as for the nonclassified case, except that in this case the crystal density distribution varies in the form of a traveling wave. As the crystal area (second moment) increases above its steady state value ($Z_2 = 1.0$), the supersaturation level is reduced below its steady state value, resulting in a reduced amount of generated nuclei. As the crystals grow and product is removed from the crystallizer, the crystal area will decrease to below its steady state value owing to the previous period of reduced nucleation. This will then lead to increasing supersaturation and an excessive amount of nucleation. As these nuclei grow to a dominant particle size, the crystal area will increase to greater than steady state values, and the cycle will start again.

The limit cycle which results for the classified product case is somewhat different from that described for nonclassified operation (*11*).

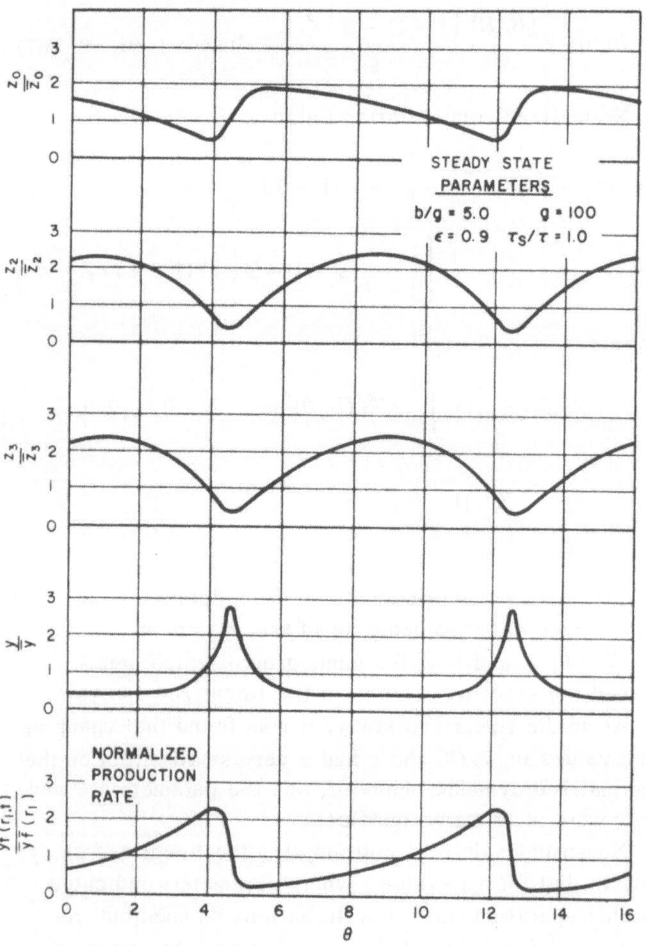

Fig. 4. Typical classified limit cycle.

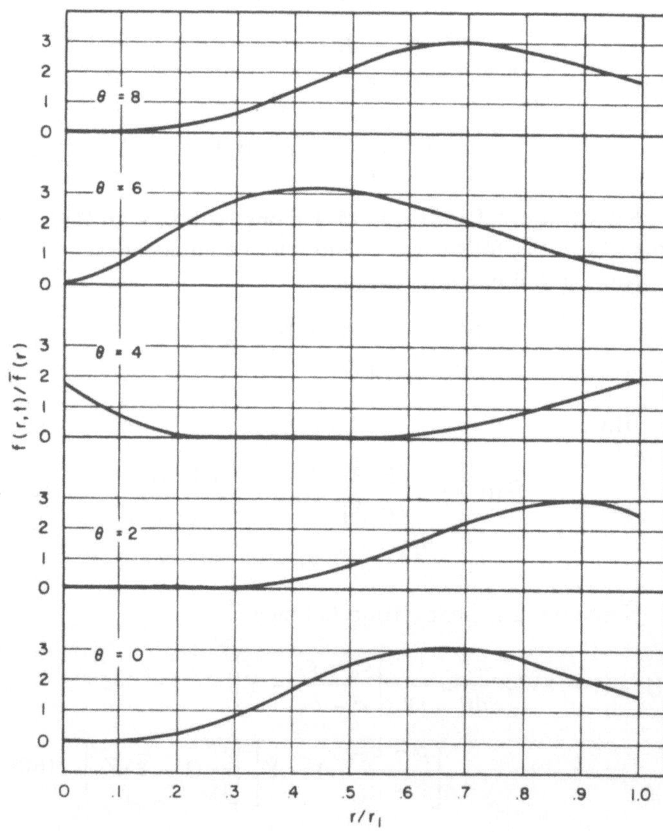

Fig. 5. Particle density fluctuations over a limit cycle.

1. The product is of constant size, but the rate of production varies.

2. The solid content of the crystallizer magma is not constant; the third moment μ_3 fluctuates throughout the cycle, and the size of its normalized oscillation is about the same as that for the second moment μ_2. Also, the fluctuations of the second and third moments are in phase with each other.

3. The cycle time is longer for classified operation than for mixed product removal (four to eight times the solids residence time).

The extent of the fluctuations and the cycle time are plotted as a function of b/g in Figure 6 for a particular voidage ($\bar{\varepsilon} = 0.9$). The similarity of the maximum and minimum values of the normalized second and third moments and of the normalized production rate should be noticed. This is due to the fact that all the moments strongly depend on the value of the density of the size distribution near the product size r_1.

SUMMARY AND CONCLUSIONS

An analysis of a special type of classified crystallizer was performed. This special model of a classified product crystallizer involved two rather strong simplifying assumptions. One is that the crystallizer is completely mixed, which is not always applicable, and the

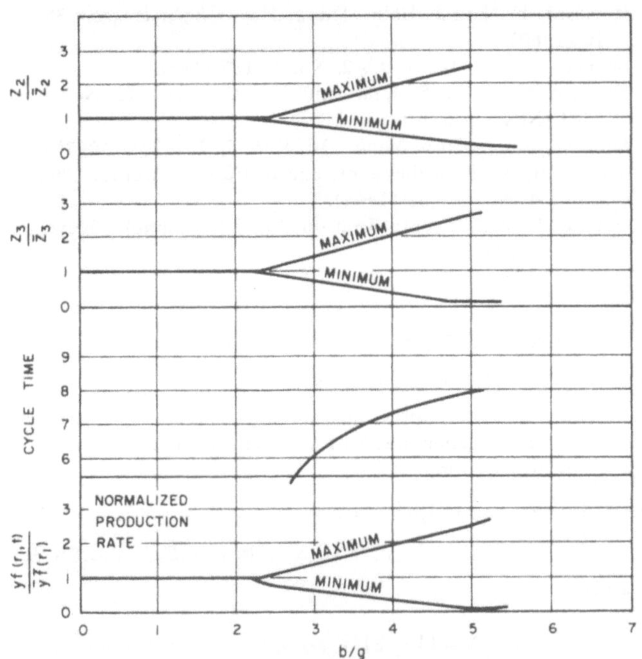

Fig. 6. Classified limit cycle characteristics as a function of b/g.

second is that the classifier is ideal and instantly removes all crystals reaching a critical size r_c. Both of these simplifying assumptions were discussed in a previous paper (*3*, *11*).

Within the framework of these commonly made simplifying assumptions, the analysis is rather rigorous and led to the following conclusions.

Similar to the mixed crystallizer, the classified crystallizer may exhibit self-induced instability leading to a cyclic behavior. The difference is that in a mixed outlet crystallizer the size of the product fluctuates while the production rate remains constant, whereas for a classified product crystallizer the product size remains constant and the production rate fluctuates as well as the size distribution and free volume inside the vessel.

The cycle period is rather long, about six to eight times the solid drawdown time, as compared with three to six times for the mixed outlet case. These long cyclic trends might, therefore, be easily masked by changes made by the operator. Depending upon the control strategy of the operator, he might either dampen or amplify the cycles, but in most cases he has a higher chance to amplify them. If the cycling has a low amplitude, it is probably best to leave it alone.

Similar to the mixed case, the source of this unstable behavior is the highly nonlinear nature of the dependance of nucleation rate on supersaturation. Especially when growing large crystals, one has to operate very close to the so-called *metastable* range. A slight perturbation in supersaturation will cause a shower of nuclei which will then grow and after a certain delay increase the area for solute deposition, thereby reducing the supersaturation below the metastable limit. Nucleation will practically stop until the available surface decreases and the concentration increases again. It is this time delay between nucleation of a particle and the time it is large enough to affect the deposition rate which, together with a nonlinear nature of $B(c)$, provides the driving force for the periodic behavior.

Again similar to the mixed outlet case, the dominant parameter affecting cyclic behavior is b/g, the nucleation to growth sensitivity ratio. However, the critical value is a magnitude lower, approximately two, as compared with twenty for the mixed outlet. This coupled with the fact that the nucleation rate must be lower for the classified outlet would indicate that the classified outlet crystallizer should tend much more toward unstable behavior. This seems to be in good agreement with general experience. The physical reason for this increased tendency toward instability is due to the fact that in the mixed outlet case any increased nucleation due to a perturbation is partially removed with the product while the crystals are small and do not yet affect the level of supersaturation.

Seeding the feed has a strong stabilizing action very similar to the mixed outlet case.

Even though this simplified model will not necessarily predict the behavior of a real crystallizer in a quantitative way, it is thought that it should preduct the basic trends correctly. It might, therefore, be used to formulate correct control strategy for a manual operator or for developing methods of feedback control. The discussion of these problems is outside the scope of this paper and might be the subject of a future paper.

ACKNOWLEDGMENT

This work was supported by the National Science Foundation under Grant GK-943. Some of this work is part of the research carried out by one of the authors (Martin Sherwin) at the City University of New York in partial fulfillment of the requirement of the degree of doctor of philosophy.

NOTATION

b = dimensionless nucleation sensitivity group
$B(c)$ = nucleation rate per volume of solution
c = solute concentration in the crystallizer
c_0 = solute concentration in the feed
c_k = metastable concentration
c_s = solubility concentration of the solute
$f(r, t)\,dr$ = number of crystals per unit volume having radii in the range r, $r + dr$ at time t
$f'(r, t)$ = crystal density perturbation
(r, t) = normalized crystal density $f/\overline{f}$
(r, t) = normalized crystal density perturbation
$F(t)$ = fraction of particles residing in a vessel for a period of time less than t
g = dimensionless growth sensitivity group

$G(c)$ = crystal growth rate dependence on concentration
$h(\theta)$ = normalized feed concentration
$j(\theta)$ = normalized feed rate
k = crystal shape factor
k_1 = constant for isothermal operations
k_n = constant
ℓ = classified case, normalized crystal size r/r_1
m = exponent in metastable nucleation function
$p(\theta)$ = dimensionless feed rate for stability analysis
$q(\theta)$ = dimensionless feed concentration for stability analysis
r = characteristic radius of the crystal
r_c = nuclei trap cutoff size
r_0 = nuclei characteristic radius
r_1 = product size for the classified product crystallizer
t = time
V = crystallizer working volume
$W(r)\,dr$ = crystal weight per volume having radii in the range r, $r + dr$
Y = normalized supersaturation
Y' = dimensionless concentration perturbation
Z_n = normalized moments
Z_n' = dimensionless moment perturbations

GREEK LETTERS

ε = fractional volume of solution
ε_p = fractional volume of solution in the outlet of a classified produce crystallizer
ζ = ratio of seed rate to nucleation
θ = dimensionless time t/τ
μ_n = n^{th} moment of $f(r)$
ρ = crystal density
τ = crystallizer drawdown time V/ω
τ_s = solids drawdown time
τ = drawdown time based on flow to the nuclei trap
ϕ = crystal growth rate dependence on r
ω = volumetric feed and/or withdrawal rate

SUPERSCRIPTS

$\bar{}$ = steady state values
$'$ = perturbation from steady state
$\hat{}$ = Laplace transform

LITERATURE CITED

1. Bransom, S. H., *Brit. Chem. Eng.*, 838 (1960).
2. Crocco, N., and S. Cheng, "Rocket Instability," Butterworths, England (1956).
3. Han, C. D., and R. Shinnar, *AIChE J.*, **14**, No. 4, 612 (1968).
4. Hulburt, H. M., and S. Katz, *Chem. Eng. Sci.*, **19**, 555 (1964).
5. McCabe, W. L., and R. P. Stevens, *Chem. Eng. Progr.*, **47**, No. 4, 168 (1951).
6. Miller, P., and W. C. Saeman, *ibid.*, **43**, No. 12, 667 (1947).
7. Murray, D. C., and M. A. Larson, *AIChE J.*, **11**, No. 4, 728 (1965).
8. Randolph, A. D., and M. A. Larson, *ibid.*, **8**, No. 5, 639 (1962).
9. Rumford, F., and J. Bain, *Trans. Inst. Chem. Engrs.*, **38**, 10 (1960).
10. Saeman, W. C., *AIChE J.*, **2**, No. 1, 107 (1956).
11. Sherwin, M. B., R. Shinnar, and S. Katz, *ibid.*, **13**, No. 6, 1141 (Nov., 1967).
12. Shinnar, R., *J. Fluid Mech.*, **10**, Part 2, 259–275 (1961).
13. Tanimoto, A., K. Kobayashi, and S. Fujita, *Interna. Chem. Eng.*, **4**, No. 1, 153 (1964).
14. Van Hook, A., "Crystallization," p. 94, Reinhold, New York (1961).
15. *Ibid.*, p. 13.

APPENDIX I: STABILITY ANALYSIS—CLASSIFIED PRODUCT CASE

The characteristic matrix for this case is given by Equation (38), and the denominator of any transfer function will have the following form:

$$A_5S^5 + A_4S^4 + A_3S^3 + A_2S^2 + A_1S + A_0 + e^{-4S}(B_4S^4 + B_3S^3 + B_2S^2 + B_1S + B_0)$$

If we let

$$N = (1 - \bar{\varepsilon})/\bar{\varepsilon}$$
$$C = (b - g)$$
$$P = g + [\tau_s/\tau - (1 - \bar{\varepsilon})]/\bar{\varepsilon}$$

these coefficients work out to be

$$A_5 = 1$$
$$A_4 = P \qquad B_4 = -N$$
$$A_3 = 0 \qquad B_3 = -PN - \tfrac{3}{4}(N + C)$$
$$A_2 = 0 \qquad B_2 = -\tfrac{3}{4}NP - \tfrac{3}{8}(N + C)$$
$$A_1 = \tfrac{3}{32}(N + C) \qquad B_1 = -\tfrac{3}{8}NP - \tfrac{3}{32}(N + C)$$
$$A_0 = \tfrac{3}{32}NP \qquad B_0 = -\tfrac{3}{32}NP$$

NYQUIST DIAGRAMS

At first we attempted to work with only the denominator of the transfer function. It was found that when this was done the number of roots on the real side of the S plane was always two more than an actual solution for the roots indicated.

We found that this was due to the fact that a factor of S^4 should have been canceled from the numerator and denominator. This becomes apparent if e^{-4S} is expanded in a power series

$$e^{-4S} = 1 - 4S + \frac{(4S)^2}{2!} - \frac{(4S)^3}{3!} + \ldots$$

and the coefficients of a new polynomial of the form

$$P = c_nS^n + c_{n-1}S^{n-1} + \ldots c_1S + c_0$$

are evaluated. The first four terms c_0, c_1, c_2, and c_3 are identically zero in both the numerator and denominator of any transfer function. (This was not true when $r_0 \neq 0$.)

When we take this into account, the contribution to the rotation of the argument when we integrate over the infinite semicircle will come out to be π instead of 5π which results if S^4 is not canceled.

The drawing of a Nyquist diagram for a transfer function containing transcendental terms results in a curve containing

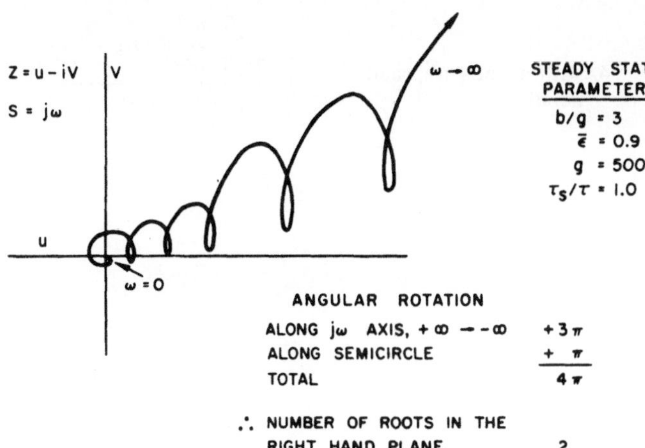

Fig. 7. Nyquist diagram.

many loops, most of which do not contribute to the net angular variation. An example of a typical result is shown in Figure 7. The number of roots will always be multiples of two unless there are roots which lie on the real axis of the S plane.

CROSSOVER POINTS

If in Equation (39) we allow $S = j\omega$, expansion of the terms results in an expression for which both the real and complex terms equal zero:

$$j\{A_5\omega^5 - A_3\omega^3 + A_1\omega - [\cos(4\omega)]B_3\omega^3 + [\cos(4\omega)]B_1\omega - [\sin(4\omega)]B_4\omega^4 + [\sin(4\omega)]B_2\omega^2 - [\sin(4\omega)]B_0\} = 0$$

$$\{A_4\omega^4 - A_2\omega^2 + A_0 + [\cos(4\omega)]B_4\omega^4 - [\cos(4\omega)]B_2\omega^2 + [\cos(4\omega)B_0 - [\sin(4\omega)]B_3\omega^3 + [\sin(4\omega)]B_1\omega\} = 0$$

If we fix all the steady state parameters except b/g, we can numerically investigate the b/g-ω plane to determine at what points the above equations hold. Only positive values for b/g and ω need be considered. Solutions to these equations will indicate that at these specific steady values a root will lie on the imaginary axis at $+j\omega$ for the particular value of b/g. Then this root will either lie to the left (stable) of the imaginary axis or to the right of the imaginary axis (unstable) for lower or upper values of b/g. In other words we know this is a crossover point for the root, but this treatment will not tell us in which direction the root is travelling with changes in b/g.

This type of study results in a series of harmonic solutions for b/g and ω.

For instance, at the steady state parameters

$$\bar{\varepsilon} = 0.0 \quad g = 100 \quad \tau_s/\tau = 1.0$$

the following crossover points were found:

b/g	ω
2.3	1.0
4.8	1.7
7.0	4.3

To determine the number of roots in right hand plane in the regions of b/g

0–2.3
2.3–4.8
4.8–7.0

it was necessary to specify the steady state parameters including b/g and to let $s = u + iv$ in Equation (39). This again will lead to two equations, both equaling zero. Then the u-v plane is searched numerically to locate the poles. For the above case there were no roots in the right-hand plane for values of b/g less than 2.3. There were two roots in the right-hand plane for values of b/g between 2.3 and 4.8 and four roots for values between 4.8 and 7.0. From this study it became obvious that the lowest crossover point was the stability limit for the system.

APPENDIX 2: STEADY STATE BEHAVIOR WITH MIXED PRODUCT WITHDRAWAL

For convenience, the equations of the steady state size distribution of the mixed crystallizer with no classification device are given below for comparison.

With the same assumption outlined in this paper, the particle balance can be written as

$$\frac{\partial f(r,t)}{\partial t} + G\frac{\partial f(r,t)}{\partial r} = \varepsilon B(c)\,\delta(r) - \frac{\omega}{V} f(r,t)$$

The steady state solution of this equation for the particle density is

$$\bar{f} = \frac{\bar{\varepsilon}\bar{B}}{\bar{G}} e^{-\frac{\bar{\omega}}{V\bar{G}}r}$$

where the superscript bar indicates steady state values.

The moments can be calculated by straightforward calculation of

$$\bar{\mu}_n = \int_0^\infty \bar{f}(r)\, r^n\, dr$$

giving

$$\bar{\mu}_0 = \frac{\bar{\varepsilon}\bar{B}}{\bar{G}}(\tau\bar{G})$$

$$\bar{\mu}_1 = \frac{\bar{\varepsilon}\bar{B}}{\bar{G}}(\tau\bar{G})^2$$

$$\bar{\mu}_2 = \frac{2\bar{\varepsilon}\bar{B}}{\bar{G}}(\tau\bar{G})^3$$

$$\bar{\mu}_3 = \frac{6\bar{\varepsilon}\bar{B}}{G}(\tau\bar{G})^4$$

$$\bar{\mu}_4 = \frac{24\bar{\varepsilon}\bar{B}}{G}(\tau\bar{G})^5$$

where

$$\tau = \frac{V}{\omega}$$

APPENDIX 3: SUMMARY OF THE FORM OF b/g FOR VARIOUS OPERATIONS

Mode	b/g
1. Normal operation no fines trap or seeding	$\frac{\bar{G}}{\bar{B}}\frac{dB}{dG}(\bar{c})$
2. With a fines trap	$\frac{\bar{G}}{\bar{B}}\frac{dB}{dG}(\bar{c}) + \frac{r_c}{\tau'\bar{G}}$

3. With seeding $$\frac{\dfrac{\bar{G}}{\bar{B}}\dfrac{dB}{dG}(\bar{c})}{1+\delta}$$

4. With a fines trap and seeding $$\frac{\dfrac{\bar{G}}{\bar{B}}\dfrac{dB}{dG}(\bar{c})+\dfrac{r_c}{\tau' G}}{1+\delta}$$

CRYSTALLIZATION OF PURE SILICON FROM MOLTEN ALUMINUM

L. M. Litz
S. A. Ring
and
R. A. Mercuri

A simple thermal convection apparatus, with no moving parts, was developed to carry out the continuous recrystallization of commercial silicon from molten aluminum. At temperatures above the eutectic temperature, this process was found to produce silicon of 99.95+ % purity in a single stage.

Recrystallization, that is, purification by solution of an impure material in a suitable solvent and subsequently precipitating a purer phase by evaporation of some of the solvent or by lowering the temperature of the solution, is one of the oldest chemical engineering operations. However, because of the technical problems involved, its application to metallic systems has not been generally realized. Such considerations as operating at fairly elevated temperatures with highly corrosive solutions, the associated problem of working in noncontaminating materials of construction, and the need for adequate techniques for material transfer and product recovery have greatly limited exploration in this area.

SOLVENT SELECTION

In the following paragraphs an approach will be described which permitted considerable upgrading of metallic silicon by recrystallization from molten aluminum solutions. As summarized in a paper by von Wartenberg (*1*), a number of previous investigators had rather unsuccessfully probed the problem of purifying silicon by recrystallization from aluminum and several other possible solvent metals such as tin, zinc, silver, and an indium-antimony alloy. As will be developed below, an important consideration which was not generally taken into account by these previous workers is the basic fact that crystals formed above the eutectic point are typically purer than those which come out in the eutectic, where all components enter the solid phase.

Aluminum was selected by us as an appropriate solvent phase on the basis of several factors. Firstly, it is probably the most universal solvent for other metals. Therefore, it very well meets the primary requirement of being a good solvent for the impurities which are to be removed. Its comparatively low melting point (652°C.) allowed the use of readily available furnace components. The fact that it does not form compounds with silicon satisfies another prime requisite. As shown in Figure 1, the phase in equilibrium with the liquids is pure silicon.

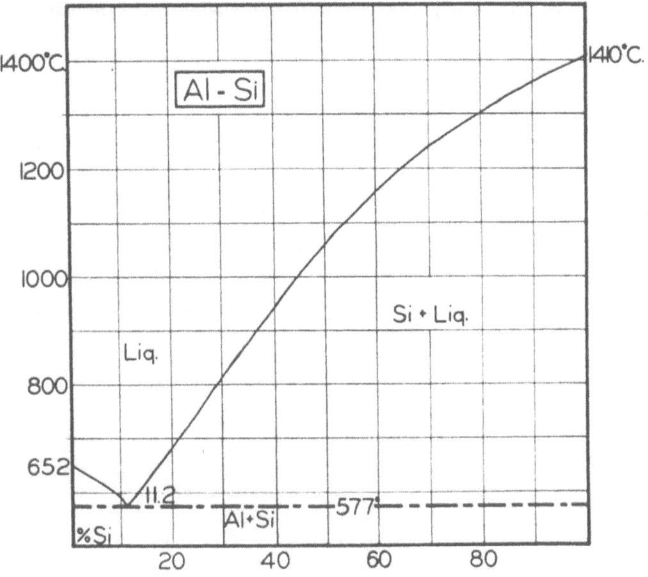

Fig. 1. Aluminum-silicon phase diagram.

Union Carbide Research Institute, Tarrytown, New York. S. A. Ring is with the Lockheed Aircraft Corporation, Sunnyvale, California. R. A. Mercuri is with the Union Carbide Technical Center, Cleveland, Ohio.

Another useful fact is that it is available commercially in high purity at a reasonable cost. Also, residual aluminum adhering to the purified silicon is easily removed by solution in common solvents. Finally, it can be contained in graphite without excessive reaction up to at least 900°C.

PROCESS THEORY

An operational approach was developed (*2*) in this work which not only resulted in the desired formation of the product above the eutectic temperature, but which also permitted continuous operation without shutting down for either addition of fresh crude silicon or withdrawal of the refined product. It had the further significant advantage that no pumps or moving seals were required to operate at these high temperatures in this extremely corrosive environment.

The basic principle of the system design is that of the thermal convection loop. That is, the driving force for circulation of fluids in the system is the density difference generated by keeping one vertical portion at a different temperature than another. The thermal convection loop has been used for many years to test materials in liquid metal and molten salt systems, but to the authors' knowledge it has not been used for recrystallization refining prior to this work.

Figure 2 is a drawing which may be used to illustrate the operating features. It shows a loop generated by connecting two vertical pipes by horizontal upper and lower transfer pipes. This loop is filled with sufficient liquid to permit flow through the upper transfer pipe. Then, if one vertical side is held at a temperature above the other, giving rise to a density difference between the two vertical legs, gravity will cause the higher density liquid to flow down and through the lower transfer pipe and the lighter liquid to flow up and across in the upper transfer pipe. Thus, in Figure 2, for $T_1 > T_2$, the flow is in the direction shown by the arrows. The rate of flow is, of course, dependent on the average density difference between the two sides and the flow resistance along the flow path.

Now if, as shown, a solid solute is placed in the hotter leg of the loop, it will dissolve in the solvent liquid. Then, if the temperature on the other side is below the liquidus temperature of this solution, the solid phase in equilibrium with the solution at this lower temperature will tend to precipitate. Purification is obtained as long as the impurity atoms do not coprecipitate with the bulk solute. The solution, saturated with the primary solute at T_2, returns through the lower transfer pipe to the hotter side, where it dissolves more solute and the cycle is repeated. To prevent crystallization in the transfer arms, it is necessary that $T_3 \geq T_1$ and $T_4 \geq T_2$.

EXPERIMENTAL

APPARATUS

The basic structure of the unit used for many of the experiments in which silicon was recrystallized from aluminum is shown in Figure 3. It was fabricated from Union Carbide ATJ graphite blocks machined to 2 in. I.D. vertical sections 20 in. long, and 1 in. I.D. horizontal tubes 10 in. long, with ⅝ in. thick walls in the cylindrical area. Flat, spring loaded Nichrome X tie bolts through crossbars at the top and bottom of the vertical legs held the system together. Machined flats at the interfaces provided very adequate seals. The purified silicon crystals were removed from the product leg of the loop by means of the product collection cup shown alongside the loop in Figure 3. This was simply a thin walled graphite

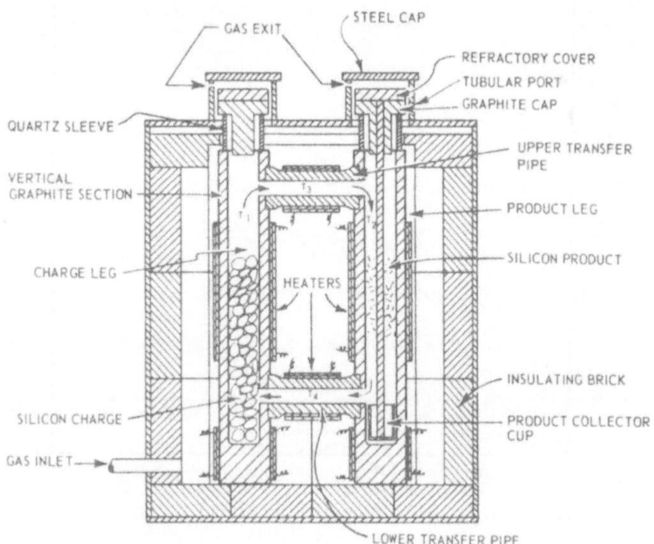

Fig. 2. Graphite loop assembly.

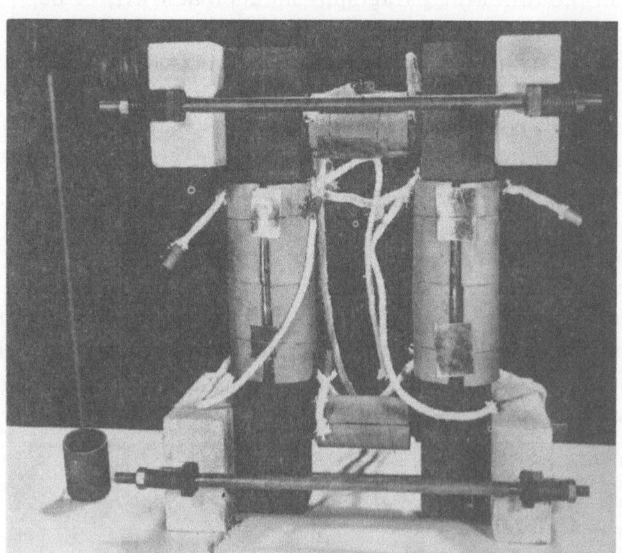

Fig. 3. Graphite loop.

cup with a perforated bottom, into the center of which was screwed a threaded ¼ in. graphite rod long enough to project above the top of the product leg. During the normal course of a run, it rested at the bottom of the product leg.

Heat was applied through a set of Hevi Duty Company Chromel wire elements supported in ceramic forms. These were available in a variety of shapes, sizes, and wattages. Semicylindrical units were fastened around the cylindrical sections, flat units on the flat faces. Each area was wired independently to an autotransformer to permit variations in the heat input as desired to attain the appropriate temperature contours. For the apparatus shown, 1,400-w. heaters on each of the vertical sections and 580 w. on each of the horizontal sections provided more than adequate power.

Since the graphite components as well as the liquid metal solution were very susceptible to oxidation, the entire unit was inserted in a gas tight sheet metal box. A vertical section of the entire assembly is shown in Figure 2. Ceramic lead-through insulators were used to bring the power leads from the heating elements through the wall of the box. The two capped ports on top of the box provided access to the two vertical sections. Short sections of quartz or refractory tubing were cemented to the top of the graphite vertical sections. These prevented spilling of molten metal on the heater elements when material was withdrawn from the operating sections. Graphite plugs, topped with disks of insulating brick, minimized the heat loss through the ports. As a further measure to reduce the heat loss from the upper areas of the vertical sections, as well as to reduce the amount of air oxidation of this area when the system was open, the vertical graphite sections terminated several inches below the port opening.

A slow argon or helium stream, fed in at the bottom of the box and out through openings in the upper ports, kept air ingress at an adequate low level. Nitrogen was not used because of the possibility of nitride formation with both silicon and aluminum. The graphite loop was well insulated from the shell by a double layer of J–M 2000 insulating brick. The temperature profile was measured with chromel-alumel thermocouples inserted in small holes at various points where possible or tied tightly to the loop. It was found that a small closed-end quartz sleeve over the end of the couple kept it from failing as a result of reactions which occurred if the couple was in direct contact with the graphite.

PROCEDURE

After the unit was preflushed with inert gas to reduce the oxygen content in the box, the heaters were energized and the temperature of the graphite loop brought up above the melting point of aluminum. For a unit such as that of Figure 3, a total of approximately 3 kg. of aluminum was inserted and melted down. The aluminum was in the form of approximately 1 in.^{2} bars, about 6 to 9 in. long obtained by sawing up standard ingots of commercial high purity aluminum. No discernible difference in purity of the silicon product was observed whether 2S aluminum (99.5% pure) or higher purity (99.98% aluminum) was used.

Commercial grade silicon lumps, typically ¼ to 1 in. in diameter, were then slowly fed into the charge side of the loop until an amount in excess of the quantity required to saturate the solution at the desired operating temperature was added. For example, when the dissolution area was run at 750°C., where the solubility is about 25%, an amount of silicon in excess of 1,000 g., perhaps of the order of 1,200 g., was used. During the initial charge period, all portions of the loop were kept as hot or hotter than the charge leg.

To start forming the purified product, the temperature of the central portion of the product leg was allowed to fall until it was from about 2° up to about 20°C. lower than the charge leg. By adjustment of the settings on the autotransformers, the desired temperatures and temperature differentials were maintained within a few degrees. As mentioned above, it was important to keep the horizontal connecting legs at a temperature in excess of any other points in the loop to insure that precipitation of silicon in these legs did not result in blockage of the circulating flow.

After approximately 1 to 1½ hr., sufficient silicon would have crystallized in the product leg to almost completely block it. At this point the steel cap would be removed from the top of the product port and the refractory cover and graphite cap taken off the product leg of the loop. By means of the graphite rod handle screwed into its base, the product collection cup was then slowly pulled up through the product leg to recover the silicon crystals residing therein. The perforations in the bottom of the cup permitted the product to be strained free of the bulk of the surrounding mother liquor. A new collection cup was then inserted in the product leg, the covers replaced, and the experiment continued.

As the given run proceeded, additional silicon would be fed into the charging leg as needed to maintain an excess of undissolved crude silicon in the loop. A graphite rod probe inserted in the charge leg provided an indication of the level of the top of the undissolved silicon. In addition, fresh aluminum was added at intervals to maintain the liquid level in the system above the top cross leg. This was made necessary because of the withdrawal of some of the mother liquor with the product.

Typically, when not collecting product, the system was put in a standby condition by holding the charge leg at the normal operating temperature (which might be between 750° and 800°C.) and by raising the temperature of the product leg above that of the charge leg. A temperature reversal of this type was also used to clear the system of blockage in the product leg if the amount of product was formed in excess of that which could be dislodged with the product collection cup.

Operating temperatures in the various experiments were of the order of 600° up to as high as 850°C. In all cases the maximum temperature difference between the charge and product leg was no more than about 20°C. In the 2 in. diameter loops the best production rates of purified silicon were approximately 60 to 70 g./hr. at temperatures of 700 deg. and higher.

A number of runs were made in a loop in which the inside diameter of the vertical legs was increased from 2 to 2½ in. and where the charge leg was maintained at 700°C. and the product leg at 750° to 755°C. Production rates up to 160 g./hr. of purified silicon were obtained. The maximum rates were realized when the collection time was held between 30 to 40 min. As a result of the larger diameter and the short collection time, restriction of circulation by accumulated products was minimized, and therefore high transfer rates were realized.

PRODUCT RECOVERY

As removed from the loop, the product collection was in the form of a cylindrical mass of platelets cemented together with the frozen aluminum solution which had been adhering to them when they were withdrawn from the loop. The cylinder diameter was essentially that of the inside of the loop, and it might be from 4 to 5 in. long and would weigh about 500 g. By inserting the cylinder in a bath of cold 9N hydrochloric acid, essentially all of the aluminum metal would be dissolved. The insoluble silicon was composed of a mixture of a very fine powder with small silicon platelets. The latter ranged from the order of a millimeter to as much as 10 to 15 mm. across. These platelets were all rather thin, with the largest being only a few tenths of a millimeter in thickness. Sieving with a 20-mesh screen submerged in water permitted the fines to be screened from the larger platelets.

Spectroscopic analysis showed that the fines were of the same order of magnitude in purity as the original silicon, whereas the coarser platelets were at least an order of magnitude more pure. This is obviously a consequence of the coprecipitation of the impurity components with the silicon as the aluminum frozen at the eutectic temperature. Based on chemical analysis and the material balance, it would appear that the fines included not only the silicon precipitated at the eutectic temperature, but also most of the silicon that is in solution in the adhering mother liquor at the time it is withdrawn with the platelets. In addition, there would be included about 10% of silicon which is recrystallized in the normal fashion, but which was small enough to go through the 20-mesh screen. In one particular sequence of runs, 807 g. of purified platelets were recovered from a total product mass of slightly over 6 kg., made up of about 40% silicon and 60% aluminum. Improved techniques for removing the extra mother liquor from the crystal mass would obviously be desirable. This might include centrifugation of the mass while still above the melting point of the aluminum solution.

The starting material contained from 1 to 3% iron with small amounts of aluminum, calcium, manganese, magnesium, and copper and trace amounts of boron, sodium, titanium, and chromium. The typical product of a single recrystallization of this crude silicon from molten aluminum by the procedure described above was spectroscopically free of all impurities except aluminum and copper. The copper content of typical platelets was from 2 to 5 p.p.m. and occasionally lower. Typical aluminum contents ranged from 200 to 700 p.p.m. Since selected single crystals from a given batch analyzed as low as 100 p.p.m. aluminum, it is assumed that the higher aluminum content normally observed was the result of occlusions between particles which were inaccessible in the washing procedure.

A small fraction of the platelets from any given run were obviously single crystals. They had a very peculiar growth habit in that they were almost optically flat on one surface, while the other side was strongly ribbed with a treelike structure. These plates were diamond shaped with acute angles of 60 deg. The ribbed face had a raised rib running on the long diagonal of the diamond with angular branches at 30 deg. to this central rib and parallel to two of the edges of the diamond. Reflection of a light beam from the ribbed surface gave coherent reflectance typical of single crystals. Laue x-ray diffraction patterns taken normal and at 45 deg. to the crystal face showed sixfold and fourfold symmetries characteristic of the 111 and 100 plane of silicon, respectively. The plane of the plates is therefore the 111 plane. This is in agreement with the 60 deg. angle between the ribs on the crystal surface.

Polycrystalline ingots were pulled by the Czochralski technique from many of the batches of silicon platelets to determine their characteristics with regard to their potential application for semiconductors use. The aluminum rapidly segregated leaving a resistance plateau of the order of 0.1 ohm-cm., p type, indicative of about 6 p.p.m. of boron. Small amounts of scum, apparently due to oxide and silicon carbide impurities, made the pulling of single crystals from these materials difficult.

LITERATURE CITED

1. von Wartenburg, H., *Z. Anorg. Allgem. Chem.*, **265**, 186 (1951).
2. Litz, L. M., S. A. Ring, and R. A. Mercuri, *U.S. Pat. 3,097,068* (July 9, 1963).

GROWTH OF ORIENTED SINGLE CRYSTALS FROM LOW MELTING METALS

H. J. Jensen
M. J. Murtha
and
George Burnet

A method (casting-bonding) has been developed for the production of large single crystals of controlled shape and a lattice orientation the same as that of a small seed crystal. Molten metal is cast in a mold, at the bottom of which is located a seed crystal held under a thermal gradient so that the upper portion is molten and the lower solid. Unidirectional cooling causes crystallization to begin at the seed. A bond forms at the seed-feed metal interface, and the lattice orientation of the seed continues throughout the entire crystal. The method has been used to grow oriented single crystals of bismuth, zinc, tin, lead, cadmium, and aluminum, some as large as 3 in. in diameter by 14 in. long.

Interest in metal single crystals began many years ago with research on metal deformation properties. Since that time, the use of single crystals in the laboratory has greatly increased with work on the solid state of matter. Frequently, the crystals must be of a specific lattice orientation for determination of physical and mechanical properties in the various crystallographic directions. An urgent need for such oriented single crystals led to the development at the Ames Laboratory of a method for producing large single crystals of controlled shape and orientation from small seed crystals of the desired final orientation. The technique has been given the name *casting-bonding.*

Crystal growth can proceed by one of four general methods: growth from the melt, growth from the solid state, growth from the vapor phase, and growth from solution. For research purposes, crystals are usually grown from the melt or solid state. Table 1 lists the common methods used.

Production of single crystals from the solid state is complicated and difficult to control but is often used to grow crystals of high melting metals. The strain-anneal and secondary crystallization methods allow a few nuclei or one nucleus to grow in a deformed crystal matrix. Controlled deformation and a precise annealing procedure are required. The crystals are normally grown from thin continuous strips of metal. Controlled cooling or thermal cycling through a phase transformation point will produce single crystals of some metals, but specifically oriented single crystals cannot be obtained by using this method.

The most reliable methods are those in which the crystal is grown from the melt. In the Czochralski (*1*) method, a crystal is nucleated on a rod which is gradually withdrawn from the surface of the molten metal. In the Bridgeman (*2*) method, the molten metal in a mold is moved slowly through a temperature gradient which induces unidirectional solidification (crystal formation) within the mold.

TABLE 1. METHODS USED TO PRODUCE SINGLE CRYSTALS FROM THE MELT OR SOLID STATE

Method	Medium
Czochralski	Melt
Bridgeman	Melt
Strain-anneal	Solid state
Secondary recrystallization	Solid state
Phase transformation	Solid state

Iowa State University of Science and Technology, Ames, Iowa.

THE CASTING-BONDING METHOD

In the casting-bonding procedure for growing oriented single crystals, a single crystal seed is metallurgically bonded to a sample of the same metal as the seed. Correct execution of the procedure gives a high assurance that the orientation of the final crystal will be the same as that of the seed. Oriented single crystals of bismuth, zinc, tin, lead, cadmium, and aluminium, all relatively low melting metals, have been produced.

Oriented single crystals of all of the above metals except bismuth are produced by using a one-step casting-bonding operation. In the case of bismuth, which expands upon solidification, the rigid mold normally used must be replaced by a soft-pack mold. A polycrystalline bismuth slug is metallurgically bonded to a small oriented seed and the resulting composite soft-packed in fine alundum powder in a stainless steel crucible. A somewhat modified Bridgeman method is then used to grow a large oriented single crystal as reported by Olson (*3*) and Slonaker (*4*).

The one-step operation uses two crucibles, one for melting the feed metal and one for casting the single crystal. The crucibles are installed one above another, the one for melting on top, inside a two zone, single core, electric resistance furnace. A sectioned drawing of the overall apparatus is shown in Figure 1. Two sizes of equipment have been used, the first to produce crystals about 0.5 in. in diameter by 4 in. long and the second much larger crystals 3 in. in diameter by 14 in. long. Representative samples are shown in Figure 2.

The metal from which the crystal is to be grown is first melted in the upper crucible. At the same time the

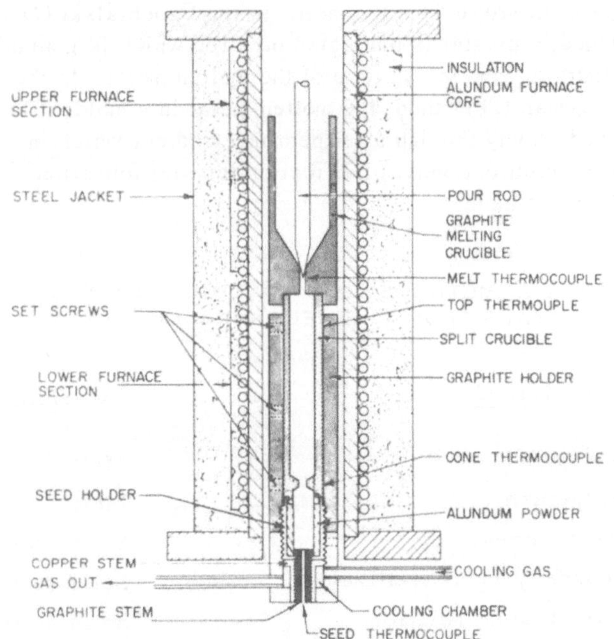

Fig. 1. Schematic drawing of the casting-bonding apparatus.

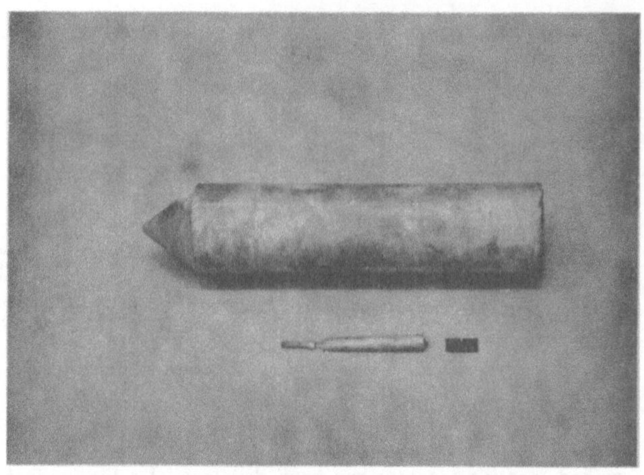

Fig. 2. Oriented zinc single crystals to show the size grown by using the casting-bonding method.

upper portion of the seed crystal, mounted at the base of the lower crucible, is heated until molten, while the lower portion is held at a temperature about 60°C. below the melting point of the metal. Precise temperature control is accomplished by using a stream of cooling air which flows through a cooling chamber near the seed holder.

The feed metal is next center poured into the casting crucible through the bottom of the melting crucible to prevent entry of contaminants such as oxide slag found on the surface of the melt. Controlled cooling then causes unidirectional solidification of the melt beginning at the solid seed-melt interface. If the seed is strain free, the lattice orientation of the seed will carry through the entire crystal. Foreign grains do form at the edge of the crystal near the seed, but they can be forced out by necking down the mold just above the seed. The proper design for both crucibles is shown in Figure 3. If the mold is not necked down, a polycrystalline slug will result. Figure 4 shows the importance of this design detail. The polycrystalline sample on the left came from a mold which was not necked down, while the oriented single crystal on the right came from a properly designed mold.

Detailed drawings of the cooling chamber and seed holder are shown in Figure 5. The graphite and copper stems are to provide good heat transfer between the air flowing through the cooling chamber and the base of the seed. The pour rod shown in Figure 3 is made from oxidized tantalum which is inert to all the molten metals used except aluminum. When the latter was to be cast, a graphite pour rod was used to assure free flow between the two crucibles.

Temperature control was accomplished by using four monitoring thermocouples located as shown in Figure 6. One couple measured the temperature at the base of the seed, a second that just below the necked down region of the casting crucible (upper portion of the seed), a third the temperature at the top of the casting crucible,

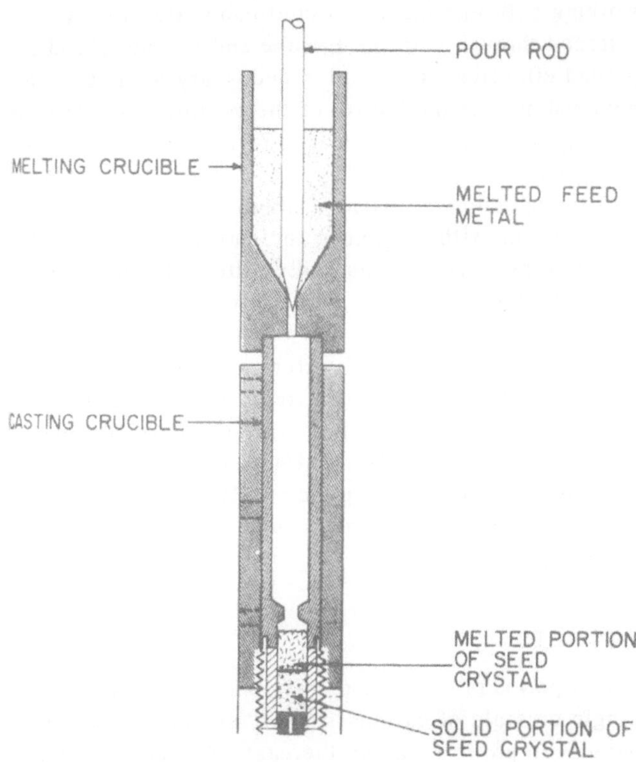

Fig. 3. Crucible design showing condition of metal before pouring.

and the fourth that in the melting crucible at the tip of the pour rod.

PREPARATION OF METALS

The first seed crystals for each metal were polycrystalline and were simply cut from pure metal slugs. Several preliminary runs were then made by using these polycrystalline seeds to determine the temperature conditions required to assure a metallurgical bond between the seed and the crystal slug.

Fig. 4. The cone section of two zinc crystals. On the left, a polycrystalline specimen from an improperly designed mold; on the right, a single crystal from a correct mold.

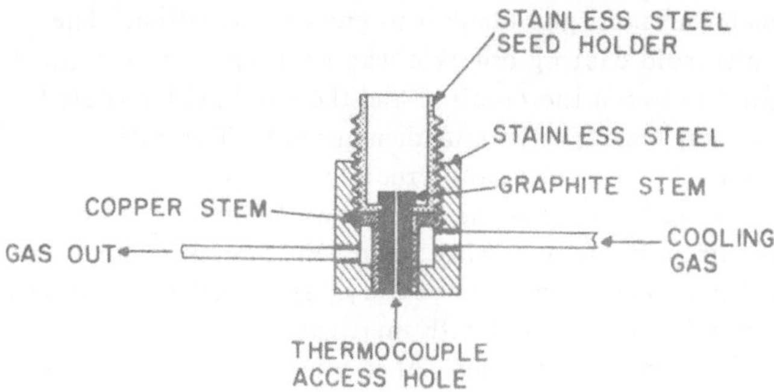

Fig. 5. Details of the seed holder and cooling chamber.

Production runs used oriented seed crystals cut free of strain from larger single crystals on a sparkcutter, After being sparkcut, the seed crystals were washed with acetone, dried, and acid etched as described by Olson (3) to remove impurities from the surface. The etched crystals were then rinsed with distilled water and stored under acetone until needed. An identical treatment was given the feed metal stock after it had been cut into small pieces for packing into the melting crucible.

PROCEDURE

The seed crystal was first positioned in the holder and soft mold 220-mesh alundum powder vibrator packed around the lower part. All operations were conducted

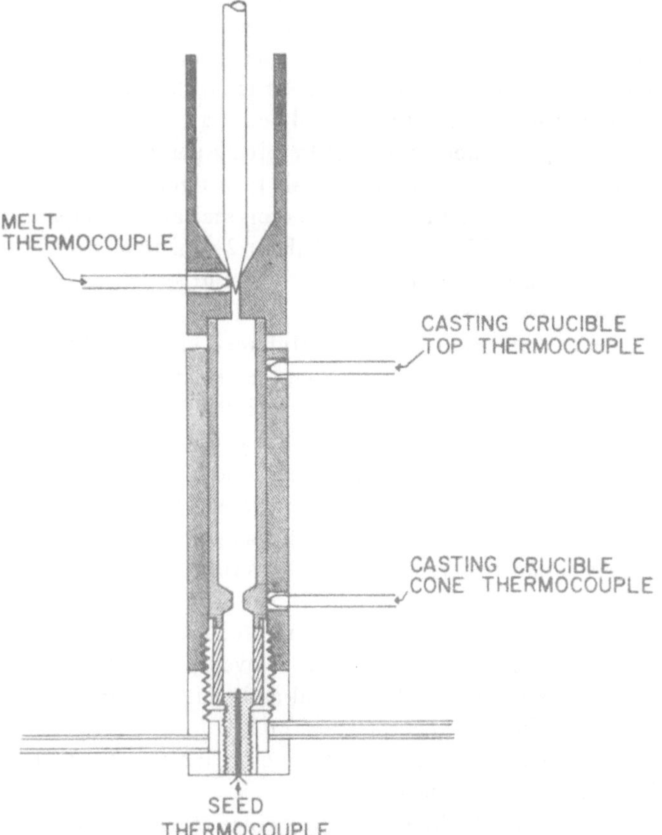

Fig. 6. Location of thermocouples.

under an inert gas blanket to prevent oxidation. The split mold casting crucible was next put in place and the joint between the crucible and the seed holder sealed against leakage with alundum cement. The outside holder for the split mold crucible was then attached to the seed holder and the set screws tightened. Care was required during assembly to prevent foreign matter from falling on the exposed seed crystal. Finally, the casting crucible was flushed with inert gas.

The furnace was then set in place over the casting crucible-cooling chamber assembly and the melting crucible placed in position on top. After the system was again flushed with inert gas, the tapered tantalum pour rod was inserted, thus sealing off the lower crucible. Feed metal pieces were packed in the melting crucible and the thermocouple leads connected to recorders.

The feed metal was next melted as rapidly as possible by applying power to both the upper and lower zones of the furnace. The temperature at the base of the seed crystal was monitored continuously. When it reached a point about 60°C. below the melting point of the crystal material, a small flow of air was initiated through the cooling chamber. Other temperatures were spot-checked through a switching arrangement.

The temperature at the neck of the casting crucible was allowed to rise to at least 50°C. but not to exceed 70°C. above the melting point of the metal to insure that the top portion of the seed crystal was molten. The temperature at the top of the casting crucible was brought to 100°C. above the melting point and that at the base of the melt crucible to at least this temperature to insure low viscosity for good pouring.

Casting was accomplished by lifting the pour rod and simultaneously increasing the air flow through the cooling chamber to maintain the seed crystal at constant temperature. After a delay of about 2 min. to assure thermal equilibrium, the air rate was further increased to remove heat through the seed crystal and cause crystallization of the melt from the solid seed upward. When the cone thermocouple indicated that the metal had crystallized to that height, power to the furnaces was reduced and finally taken to zero. The optimum freezing rate was a function of crystal size and composition. A smaller diameter crystal could be grown rapidly (8 to 10 min.) without the formation of shrinkage pipes, but larger diameter crystals had to be formed more slowly (8 to 10 hr.).

Upon reaching ambient temperature, the furnace, crucibles, and split mold were removed with extreme care so as not to jar the crystal and cause strains to develop. The crystal was then etched to reveal any grain boundaries and x-rayed to verify the single crystal structure and orientation.

Because of the higher melting point of aluminum (660°C.), it was necessary to protect the graphite and copper components of the apparatus from oxidation when working with this metal. A continuous flow of dry nitrogen through both the furnace and cooling chamber proved effective. It was also necessary to enlarge the pour rod hole at the bottom of the melting crucible from 1/8 to 5/16 in. in diameter because of the lower density of aluminum.

Although properly oriented single crystals resulted from all runs with aluminum, the surface of each of the first several crystals was badly pitted. Figure 7 shows one of the pitted crystals and a later specimen grown essentially free of pits. The pits were found to be caused largely by trapped water vapor and were eliminated by baking out all graphite components and alundum powder prior to use. Starting with ultrapure aluminum (99.999%) was also helpful. Under these conditions, good quality oriented single crystals were consistantly grown without the need to conduct the entire operation under high vacuum.

RESULTS

In a period of 4 yr., more than one-hundred oriented single crystals of various shapes, sizes, and metals have been grown by using the casting-bonding method. Reliability has been high, and only about 10% of the runs attempted have failed to produce the results desired. About one-half of the failures have occurred because a bond was not formed between the seed and the feed metal. Figure 8 shows representative crystals of the five metals for which the casting-bonding method has been used.

The integrity of the metallurgical bond between the seed and feed metal was checked on many samples by cutting the crystal axially at the bond site. In no case

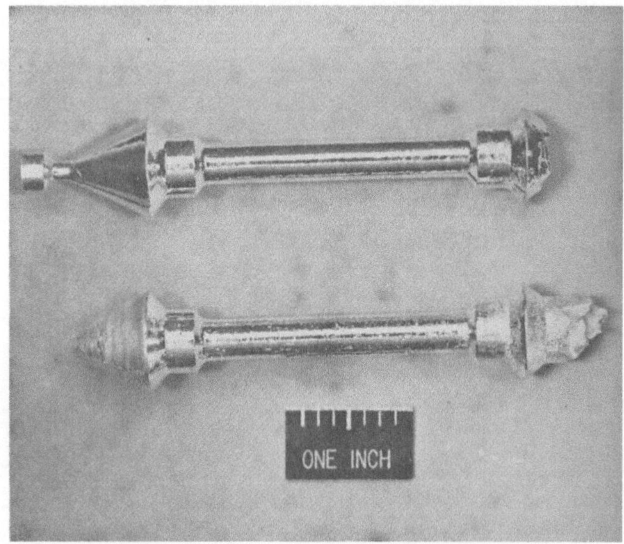

Fig. 7. Aluminum oriented single crystal tensile specimens. Pitted surface on the sample to the right was caused by moisture.

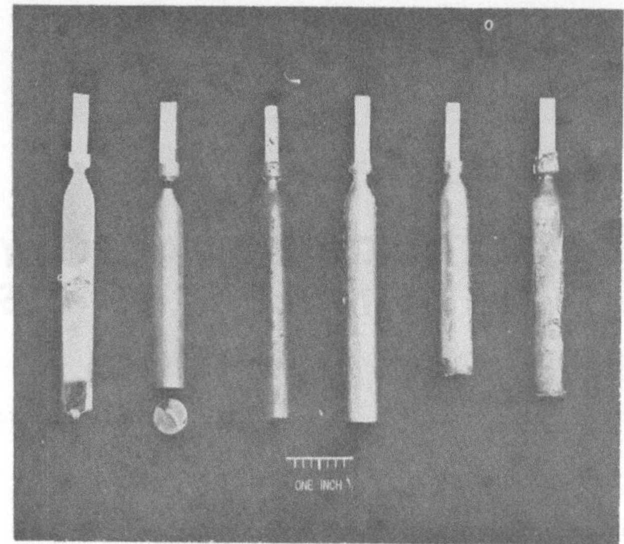

Fig. 8. Crystals grown from oriented seeds. From left to right, zinc with axial cleavage plane, zinc with radial cleavage plane, lead, tin, aluminum, and bismuth. The rough surface on the bismuth is from the soft mold.

was there evidence of a boundary or a concentration of impurities or imperfections.

In summary, the casting-bonding method is capable of growing, with an unusually high degree of reliability, shaped, oriented single crystals of several relatively low melting metals. Future work will include an investigation of crystal perfection as it relates to key operating variables.

ACKNOWLEDGMENT

Work was performed on the Ames Laboratory of the United States Atomic Energy Commission, contribution number 2251.

LITERATURE CITED

1. Czochralski, J., *Z. Physical Chem.*, **92**, 219 (1917).
2. Bridgeman, P. W., *Proc. Am. Acad. Arts Sci.*, **60**, 305 (1925).
3. Olson, E. H., *USAEC Rept IS-178* (1960).
4. Slonaker, R. E., Jr., *J. Less-Common Metals*, **7**, 165 (1964).

MELT GROWTH OF REFRACTORY OXIDE SINGLE CRYSTALS

Howard M. Dess

In the same manner that the discovery of the transistor in the 1940's triggered the start of an enormously productive era of crystal growth work on silicon, germanium, and related semiconductor materials, so did the first successful demonstration of laser action in ruby in 1960 (*3*) act as a powerful stimulus for crystal growth research and development efforts in the field of refractory oxidic materials. In both cases, extraordinary exertions by crystal growers were necessitated by the urgent demands for single crystals with unprecedented purity levels, degree of perfection, and large size. The laser situation was further complicated by the fact that many of the compositions of greatest potential interest as hosts were much higher melting (for example, melting point of ruby, 2,050°C.; $Y_3Al_5O_{12}$, YAG, 1,970°C.) than the most familiar semiconductor materials (germanium, 958.5°C.; silicon, 1,410°C.). This, in turn, relates to problems of containment and contamination of melts which in general are quite different for these two classes of materials. Thus, to a large extent, the laser crystal growth technology developed independently of the semiconductor technology, although in some instances, parallel lines of development were followed.

The general techniques available to the crystal grower may be broadly classified as melt growth, solution growth, vapor growth. A more detailed breakdown of each of these major categories is shown in Table 1 (*1*). In the laboratory, the choice of one method over another is to some extent discretionary. However, for purposes of production, the ultimate decision represents the best trade offs among a host of sometimes conflicting factors such as physicochemical properties of the material to be grown in single crystal form, performance requirements of users as reflected in quality specifications, size and shape of crystals, controllability and reproducibility of the process, and cost. In recent years, notable success with the melt pulling or Czochralski technique on an industrial scale of operations both in the semiconductor and laser fields has made this a favored approach. Historically, however, the Verneuil process originating in France just after the turn of the century represented the first practical method for producing single crystals of high melting oxides of useful size and acceptable quality on a continuing commercial basis.

TABLE 1. OUTLINE OF REPRESENTATIVE CRYSTAL GROWTH TECHNIQUES

I. Melt Growth
- A. Techniques Requiring Crucibles
  1. Melt pulling or Czochralski techniques
  2. Stockbarger, Bridgeman methods
- B. Crucible-less Methods
  1. Verneuil-type techniques
  2. Skull melting
  3. Float zone methods

II. Solution Growth
- A. 1 atmosphere, near-room temperatures.
  Crystallization from aqueous solution by solvent evaporation or cooling.
- B. 1 atmosphere, high temperatures.
  Flux growth - crystallization from molten inorganic salt solutions usually by cooling.
- C. Superatmospheric pressures, high temperatures.
  Hydrothermal methods - crystallization from aqueous systems confined in pressure vessels.

III. Vapor Growth
- A. Sublimation
- B. Chemical Vapor Deposition

Union Carbide Corporation, San Diego, California.
Howard M. Dess is with the National Lead Company, Hightstown, New Jersey.

In the following section, the material requirements necessary for the general applicability of any melt growth process are outlined. In the next sections, these criteria are then specifically related to the Verneuil and Czochralski processes which are described and compared. Finally, the applicability of these techniques to the production of specific products is discussed.

RELATIONSHIP OF MATERIAL PROPERTIES TO CHOICE OF GROWTH TECHNIQUE

In general, successful growth of single crystals via a melt process depends upon the material in question meeting the following minimal requirements:

1. Congruently melting compound.
2. Low vapor (or decomposition) pressure at temperatures above the melting point.
3. No disruptive phase changes in the solid state.

Examples of commercially important single crystal materials which cannot be produced by a melt growth process because they fail to meet one or more of the above criteria are:

$Y_3Fe_5O_{12}$ (YIG)—melts incongruently. Produced via solution growth from an inorganic flux.

Calcium carbonate—decomposition pressure of carbon dioxide exceeds 1 atm. below the melting point. Still obtained from natural sources.

Quartz—solid state phase transitions result in disruptive cracking. Grown hydrothermally at temperatures below the transition point.

A fourth requirement must be added if one of the bulk melt growth techniques is to be applicable: the melt must be compatible with available crucible materials. This means firstly that the melting point of the crucible material must be safely above the maximum melt temperatures anticipated. A 250° to 300°C. differential has been found to be a practical limit. Secondly, the crucible must be as nearly inert and unreactive as possible under the usual operating conditions. This means stability with respect to the operating atmosphere, the crucible supports, and any other contact insulation and, of course, the melt itself. Thus, for the containment of refractory oxide melts, carbon is unsuitable because it is too strongly reducing. Crucibles fabricated out of other refractory oxides are also inapplicable because of solubility effects. The range of choice finally narrows down to a few metals, the ultimate selection of which is primarily a function of melting point and reactivity. Thus, if the crystal to be grown melts below 1,500°C., platinum (melting point 1,769.3°C.) will probably prove to be a satisfactory crucible material. If the material in question melts above 1,500°C. but below about 2,200°C., then iridium (melting point 2,450°C.) is probably the best choice from the remaining noble metals (osmium and rhenium have higher melting points but are significantly more reactive than iridium, especially under oxidizing conditions; rhodium, melting at 1,960°C., offers only marginal improvement over platinum). Metals such as tungsten, molybdenum, or tantalum (melting points: 3,380°, 2,620°, 2,977°C.) although sufficiently refractory and relatively inexpensive in comparison to platinum or iridium, suffer from the disadvantage of oxidizing too easily at elevated temperatures. Thus, their utility is severely limited to those materials whose melts are stable under reducing or strictly neutral atmospheres.

THE VERNEUIL PROCESS

For those refractory oxide crystals capable of being produced in single crystal form by a melt growth technique, the Verneuil process was the favored approach until fairly recently because no crucible is required. The main features of this technique are shown schematically in Figure 1 for an oxy-hydrogen fired system. This type of setup can be used to produce single crystals whose melting points range as high as 2,300° to 2,400°C. (for example, rare earth oxides). However, plasma arc torch heat sources have also been used and in fact are preferable for materials sensitive to redox reactions, since argon is then used rather than the hydrogen/oxygen mixture. [The plasma arc torch also provides much higher temperatures than the classical gas flame torch, so that crystals with melting points well above 2,400°C. can be grown, for example, tungsten (*24*).] The powder hopper above contains the material to be grown, in finely divided form. A simple hammer tapping device or more sophisticated vibratory attachment provides the means for regulating the flow of powder from the hopper to the growth region below. Although various modifications of the gas inlet and mixing arrangements have been developed over the years, the scheme shown in Figure 1 is fairly representative. The oxygen-hydrogen flame is directed downward, inside the insulating cavity, onto the tip of a seed rod inserted from below. The bottom end of the seed rod is fastened securely in a chuck which can be programmed to raise or lower automatically at any desired rate. Temperature at the tip is adjusted by control of the hydrogen-oxygen flow rates, furnace geometry, and position of the seed. However, in all cases, what is desired is the formation of a thin, stable, molten layer on top of the seed. This condition is best judged visually and controlled during the course of a run so as to avoid either extreme of excessive liquid buildup in the cap which may lead finally to spillover, or insufficient liquid in the cap, a situation which can result in flash freezing. The main parameters most directly affecting the condition of the cap during a run are powder feed rate, heat input (that is, a function of hydrogen/oxygen flow rate and geometry of furnace insulation), and rate of withdrawal of the seed rod. During the initial stages of a run, these three factors

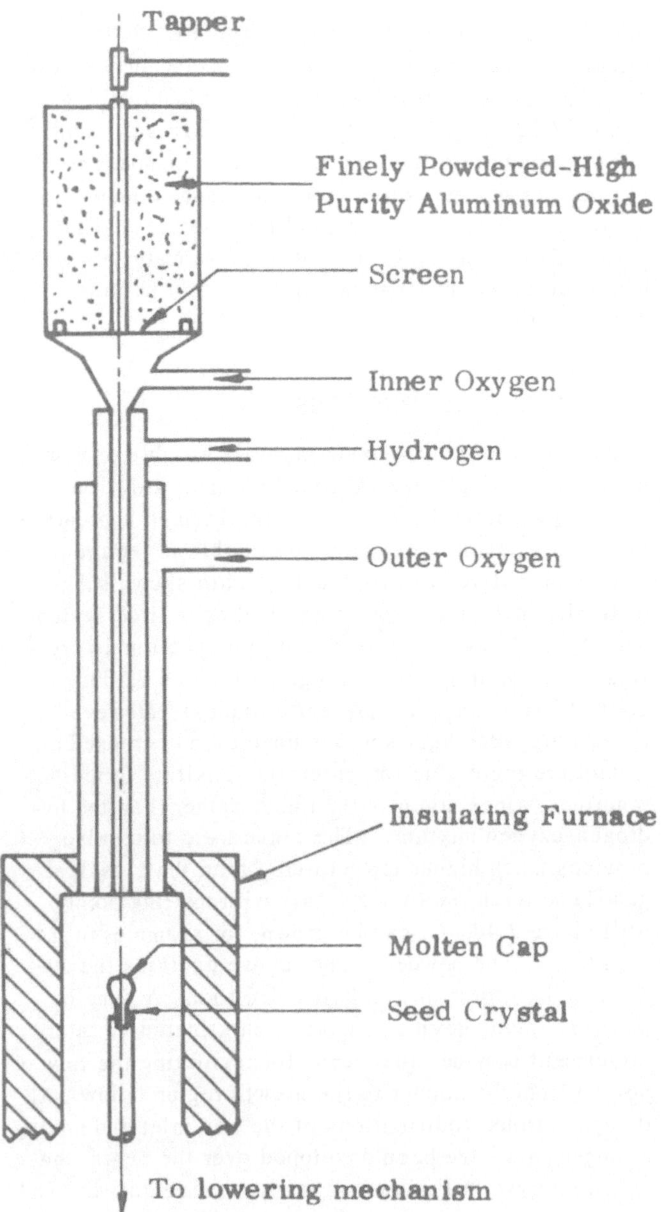

Fig. 1. Schematic diagram of Verneuil (or flame-fusion) process.

will be adjusted so as to favor widening the diameter of the seed rod to the extent desired; this may entail, for example, quite slow initial downward displacement of the seed rod. At the desired diameter, the rate of withdrawal of the seed rod will be increased so as to establish steady state conditions. However, some trimming of the gases and powder feed rate may also be required.

The specimen depicted in Figure 2 typifies the appearance of an as grown Verneuil ruby. The frosted appearance of the sides is caused by adherence of feed powder particles which miss the molten cap. The Verneuil process is still the favored method for manufacturing gem crystals such as star sapphires (see Figure 3) in which, contrary to usual crystal growth practice, a relatively insoluble dopant (titanium dioxide) is added

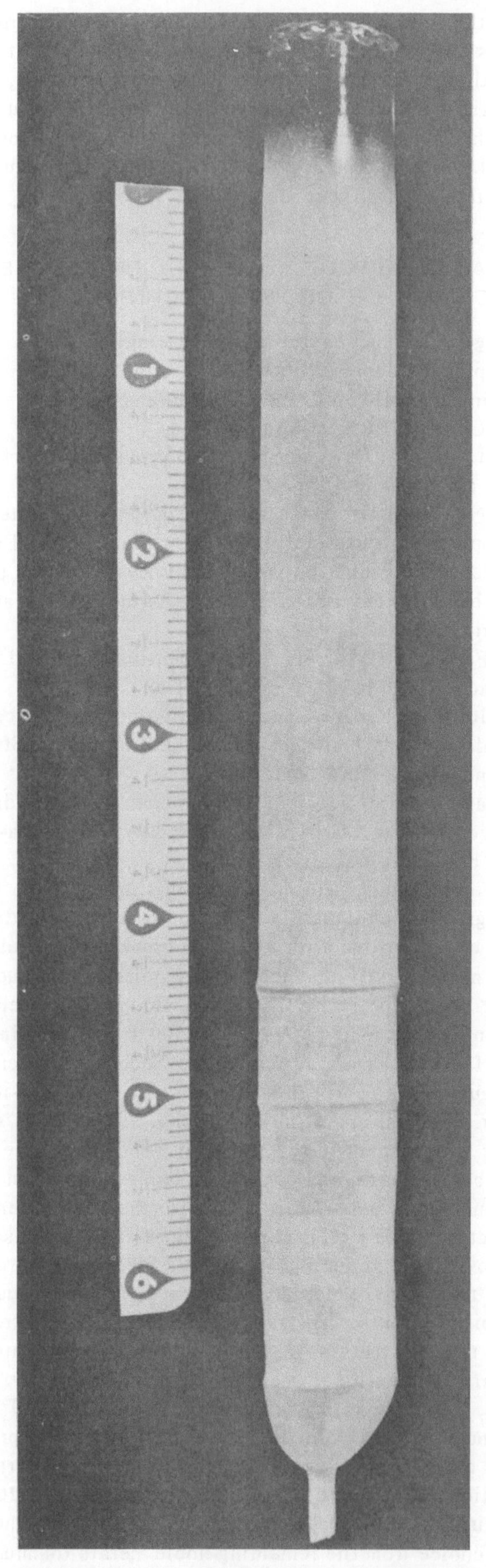

Fig. 2. Typical appearance of as grown Verneuil ruby.

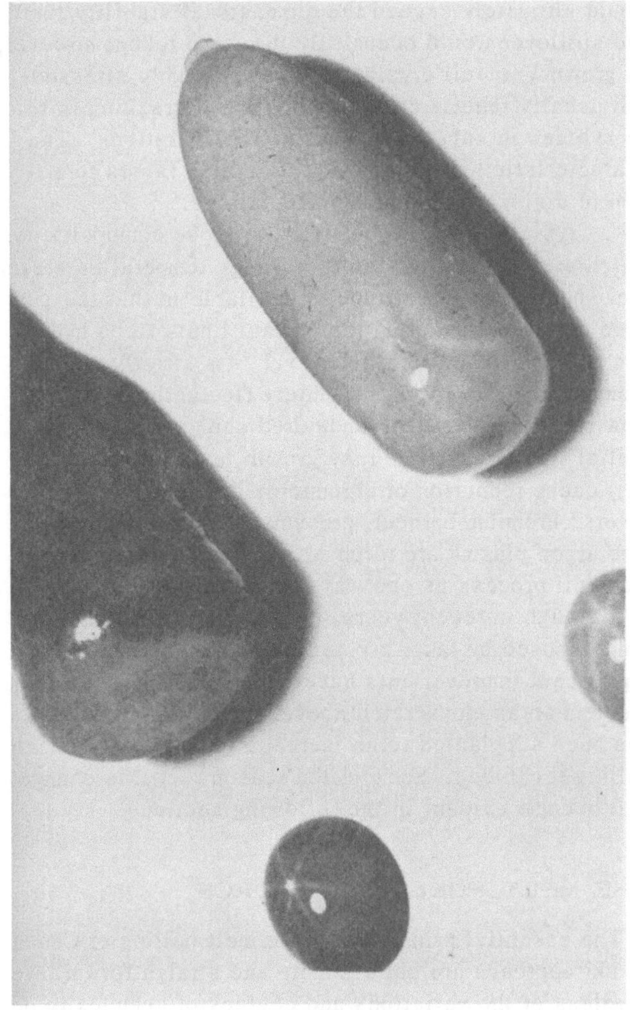

Fig. 3. Star sapphire boules and finished stones.

with the deliberate intention of causing it to precipitate out later, in the solid aluminum oxide lattice, under a carefully controlled set of annealing conditions. This crystallographically oriented precipitate of needlelike titanium dioxide crystals moreover possesses a significantly higher refractive index (2.615) than that of the host lattice (1.767), so that optical contrast between the two is quite striking. It is this difference which forms the basis for the so-called *asteriation* effect illustrated by the cabochon shaped specimens in Figure 3.

The Verneuil process is not restricted to the growth of cylindrical shaped crystals. If the seed is mounted horizontally in the furnace cavity as shown schematically in Figure 4, and rotated slowly, disk shaped crystals can be produced. Sapphire disks with diameters as large as 6 in. have been obtained in this manner. Figure 5 shows two such disks, one in the as grown condition, the other after fabrication. Still another shape variant, domes (Figure 6), can be produced by the Verneuil technique, if a horizontally mounted seed rod is slowly displaced in a horizontal direction.

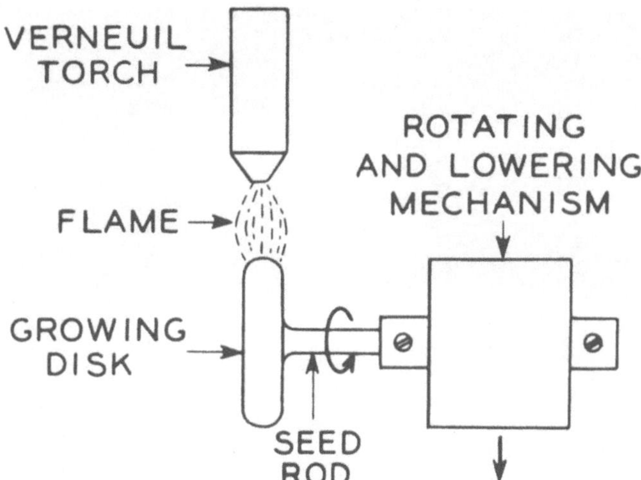

Fig. 4. Schematic representation of disk growth via the Verneuil process.

Finally, although for many years the Verneuil process has served as the only commercial source of ruby and sapphire products, the flexibility afforded by this technique is further illustrated by its adaptation to the growth of literally scores of widely varying single crystal compositions which, at the time, could not be obtained conveniently by any other method.

The preceding examples serve to demonstrate that the continuing attractions of the Verneuil process have been largely based on its versatility and relative simplicity of operation. However, because of the ultrastringent crystal quality requirements of most laser applications, it was found in the early 1960's that the previously established standards of excellence were no longer sufficient, and the Verneuil process posed certain control problems that could be directly correlated with crystal defects of various types. Some of the more obvious sources of difficulty are, unfortunately, related to operational features almost inherent in the method and therefore not readily changeable:

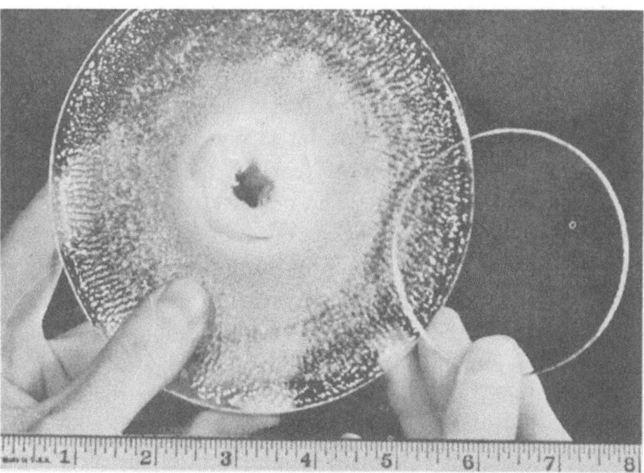

Fig. 5. As grown and finished sapphire disks.

Fig. 6. As grown sapphire dome.

1. The transport of feed powder to the molten cap is intermittent, and this factor is partly responsible for periodic temperature fluctuations in the melt and therefore also in the crystallization rate. When a dopant is present whose distribution coefficient (k = concentration of dopant in solid/ concentration of dopant in liquid) is markedly different from 1, such fluctuations in growth rate cause a corresponding fluctuation in dopant concentration readily visible to the eye as a series of closely spaced striae.

2. Another problem related to the discontinuous nature of the powder feed is that a momentary and localized oversupply of powder at one location of the molten cap may occasionally exceed the thermal capacity required for melting. The unmelted slug is then finally encased by single crystal material and appears in the finished specimen as an inclusion or scattering site. In addition, this polycrystalline inclusion may nucleate undesired orientations which propagate in the direction of growth.

3. The longitudinal thermal gradient along the crystal is quite sharp, especially in the vicinity of the cap. This is almost an essential requirement of the Verneuil process; if it were not so, the size of the liquid zone would ultimately exceed the dimensional stability limit, and spillover would occur. By the same token, however, as grown Verneuil crystals tend to be highly stressed and usually require careful annealing if cracking is to be avoided in subsequent fabrication operations. The characteristic high stress situation also favors formation of dislocations and misorientations.

4. Crystal growth is restricted to those compositions which are chemically stable in a high temperature steam atmosphere. Some latitude is available in that the gas ratio can be adjusted within certain limits to be hydrogen rich or oxygen rich. However, if a strictly neutral atmosphere is desired, momentary fluctuations in gas flow rates may result in undesired changes in redox potential (for example, in ruby growth, excess hydrogen will cause reduction of chromic oxide; excess oxygen favors chromium burnout, presumably as chromic acid). The argon plasma arc torch was adapted for use in the Verneuil process as one way around this problem.

Although in recent years, in response to the challenges posed by laser crystal quality requirements, significant improvements have been made in the various problem areas enumerated above, the Verneuil process has been supplanted to an increasing extent by the melt pulling technique. Some of the reasons for this change will become evident in the following sections.

THE MELT PULLING APPROACH

The essential principles of the melt pulling or Czochralski approach are quite simple and straightforward, regardless of the variations and degrees of complexity of experimental setup described for the many individual crystal systems studied in the past. Figure 7 illustrates in schematic fashion the key features. A melt of the substance to be grown is produced and maintained in a suitable container. A seed crystal of desired orientation is inserted into the melt. After sufficient time is allowed for thermal equilibration of the seed with the melt, pulling is commenced (that is, the seed is slowly raised in a vertical direction), and the diameter of the crystal is controlled by suitable adjustments of the pull rate and thermal environment. The crystal is usually rotated throughout the duration of the run. Types of heat sources vary with the preferences of the operator and the requirements of the crystal system. The following are illustrative: hydrogen/oxygen flame furnaces, resistance furnaces (for example, Globar, platinum resistance, etc.), and induction furnaces. As might be anticipated, the electric furnaces have proved more amenable to tight control than the gas furnaces. One type of control scheme that has performed satisfactorily with radio frequency induction furnaces is depicted in Figure 8. This method monitors and controls the power output of the furnace. Another approach is to monitor temperature at some fixed point in the system (for ex-

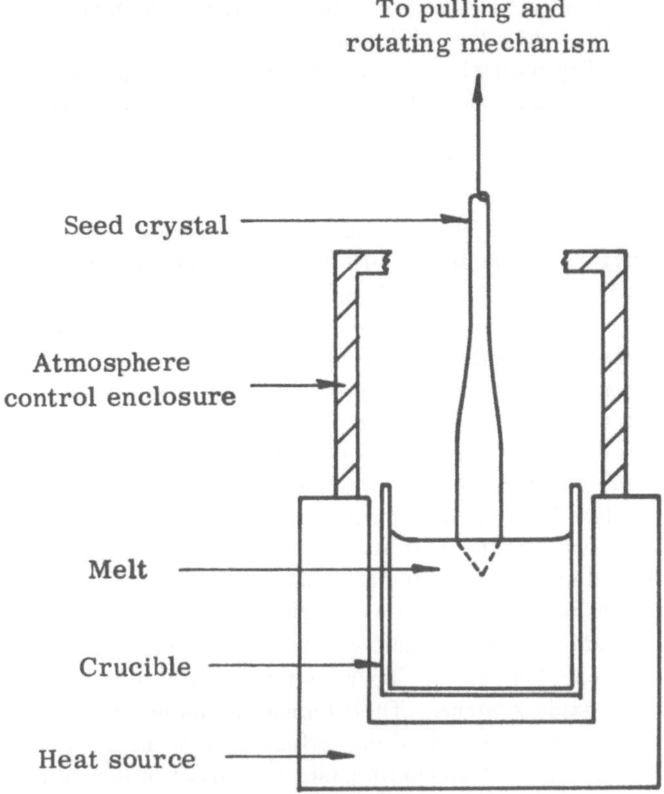

Fig. 7. Schematic representation of the Czochralski or melt pulling process.

ample, at the bottom of the crucible by means of an optical pyrometer or thermopile, or directly in the melt via a suitable thermocouple) and to relate this information to furnace power. Regardless of which specific control scheme is chosen, however, in every case the fnrnace operator is still a very important part of the system. By means of visual observations he must decide whether the crystal is growing as desired and, if not, make appropriate changes, usually in furnace power. Furthermore, because growth rates of high

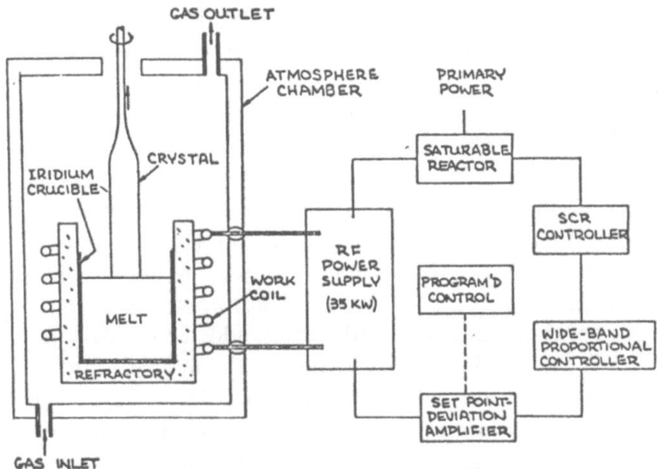

Fig. 8. One type of control scheme for a radio frequency induction furnace.

quality laser crystals are typically slow, measured in fractions of an inch per hour, the operator must actually try to anticipate potentially damaging circumstances and forestall them by taking preventive action. The only practical way this can be done is with a system whose behavior has already been well established, so that growth is a fairly routine procedure. Needless to say, methods for automatically monitoring and controlling all stages of crystal growth are being actively investigated.

The following discussion describes a fairly representative sequence of steps from start to finish of a melt pulling run. The seed is slowly lowered into a melt that has been superheated a few degrees so as to prevent flash growth when initial contact is made. The melt temperature is then gradually adjusted downward until the seed is in stable equilibrium with the melt (that is, neither expanding nor shrinking in size). Pull is then started, and simultaneously the furnace power (and melt temperature) is reduced by small step-by-step decrements over some given time period so that the crystal diameter gradually increases. However, as the diameter of the crystal becomes large relative to the diameter of the melt, it tends to behave as an ever more efficient heat sink. This is especially true for the higher melting compositions which also happen to be very transparent to radiation from the melt. Under these conditions, the dominant mechanisms for heat loss from the growth interface may finally be via radiation and conduction through the crystal, both of which are proportional to the cross-sectional area of the solid:

$$\text{Conduction losses} \propto D^2 \Delta T$$
$$\text{Radiation losses} \propto D^2 T^4$$

where D is the diameter of the crystal.* Thus, if the diameter is allowed to expand sufficiently, the crystal in effect may begin to create its own thermal environment which can soon override the external control imposed on it; that is, the diameter may begin to expand uncontrollably. Therefore, during the final stages of the widening process, it may actually be necessary to begin increasing furnace power again slowly in order to snub such a potentially disastrous occurrence. After the desired crystal diameter has been attained, very small periodic increases in furnace power may be required to maintain straightness; that is, as the crystal increases in length, it becomes a still better heat sink because the upper surfaces move into progressively cooler regions of the furnace. Finally, however, if the crystal is grown large enough so that the melt level has been reduced a significant amount, the thermal environment

*These same considerations account for the conical growth interface typically observed on many Czochralski grown crystals (see Figures 9 and 10). The melting point isotherm is actually pushed beneath the surface of the melt.

in the vicinity of the melt interface is influenced to an increasing extent by radiation from the walls of the crucible as they are slowly uncovered. This warming tendency must then be countered, if continued straight growth is desired, by periodic decreases in furnace power.

Growth parameters such as pull rates, rotation rates, thermal gradients, etc., can only be discussed here in the broadest terms. In general, that combination of growth conditions is sought which results in the highest yields of crystals of demonstrably acceptable quality levels. For laser applications, crystal quality specifications typically center on internal optical disturbances such as the following:

1. Concentration, distribution, and type of second phase scattering centers. Bubbles or voids, oxide precipitates, metallic inclusions originating from the crucible are typical examples of undesired inclusions that are responsible for inefficient laser performance.

2. Optical inhomogeneities attributed to local refractive index variations which, in turn, may be caused by nonuniform dopant concentration (or impurities, considered as unintentional dopants) or strain.

A priori selection of the best set of growth conditions that will avoid the worst of the problems noted above for a previously untried system is difficult or impossible. However, although a comprehensive theoretical analysis of the relationships between crystal quality and environmental conditions at the growth interface has not yet been carried out specifically for oxide systems, many of the concepts developed originally from experimental data and observations obtained with metallic and semiconductor systems appear equally applicable to oxides, at least in a qualitative sense. Thus, for the general case involving growth of a crystal from a melt containing a dopant whose distribution coefficient differs from 1, some of the more important aspects of internal quality may be better related to the growth situation by assuming that immediately adjacent to the growth interface there exists a thin layer of liquid (the so-called *diffusion layer*) within which the concentration of dopants may vary markedly from their level in the bulk melt. The formation of this diffusion layer is the result of a combination of circumstances: the fact that $k \neq 1$ means that some melt constituent is either preferentially rejected ($k < 1$) or absorbed ($k > 1$) by the growing lattice, and this localized deviation from bulk melt concentration values can persist because of incomplete liquid mixing at the growth interface. A detailed development of the full implications of this situation is outside the scope of this paper (references *2a*, *b*, *c*, and *12* and *13* are recommended as an introduction to pertinent literature on the subject); however, some of the major consequences are summarized below:

1. Slow growth rates and slow rotation rates favor maximum segregation of dopants.

2. Fast growth rates and fast rotation rates tend to reduce segregation of dopants.

3. The thermal gradient in the vicinity of the growth interface must be sufficiently high to suppress possible constitutional supercooling in the diffusion layer caused by compositional changes. Supercooling is extremely undesirable, since it may be the cause of quality aberrations ranging from second phase inclusions to departure from single crystal growth (that is, dendrite formation). The higher the dopant concentration, the higher must be the ratio of g/v, where g is the thermal gradient at the growth interface and v is the growth rate.

Some actual values for growth rates and rotation rates employed in the production of sapphire, ruby, or YAG illustrate both the applicability and practical limitations of the above guidelines in oxide systems. Good quality sapphire (that is, high purity single crystal alumina) can be grown from the melt at rates ranging up to 2 in./hr., depending upon the crystal diameter. As the growth rate is increased, bubbles appear in increasing numbers in characteristic periodic clusters. The appearance of such bubbles sets an upper limit on growth rates for many oxide systems. Their formation can be explained within the framework of the diffusion-layer theory if one assumes that atmospheric gases dissolved in the melt behave as dopants with a negligibly small segregation coefficient, that is, $k \ll 1$. This means that the crystal lattice continuously rejects the gaseous impurity, and concentration of this dopant tends to increase in the diffusion layer above its equilibrium value in the bulk melt. If crystal growth is sufficiently slow so that diffusion-convection processes in the melt can keep the gas concentration at the growth interface below the supersaturation level where bubble nucleation becomes energetically favorable, it will remain in solution. However, experimental evidence to date shows that the growth rate can always be increased to some level at which the gas supersaturation in the crystallization surface finally reaches a value that permits bubble nucleation. After the first bubble has formed, it can then act as a nucleation site for others. In this chain reaction fashion, a lenticular or conical shaped cluster of bubbles can be formed which roughly outlines the profile of the growth interface during that time period. Bubble formation tails off as the diffusion layer becomes depleted in gas content, but the regular distribution of bubble clusters along the length of a crystal indicates that the process of gas buildup followed by depletion can occur in near periodic fashion during a growth run.

As solid dopants are deliberately added to a melt (for example, those desired as active laser ions), growth rates must be reduced still further in order to obtain the desired degree of optical homogeneity in the crystal. Furthermore, in general, the smaller the dopant distribution coefficient compared to 1, and the higher the concentration in the melt, the slower must be the growth rate. Thus, laser quality ruby containing typically about 0. 03 wt. % chromic oxide ($k \approx 1$) (*4*) should be grown no faster than a few tenths of an inch per hour. In the case of YAG containing ~1.3 to 1.5 atomic % Nd, k_{Nd} is significantly less than 1 (*5, 6*), and very much slower growth rates are essential if the necessary internal quality is to be achieved. Growth rates in this case must be held down to a few hundredths of an inch per hour. As growth rates from doped melts are raised, crystal quality will be increasingly jeopardized by the possible occurrence of local fluctuations in dopant concentration, second phase precipitates, bubbles which tend to form at lower growth rates in doped vs. undoped melts, and dendritic growth.

The effects of rotation rate on crystals grown by a melt pulling process are not always easy to interpret, since other factors tend to be overriding and to mask or distort the results. Perhaps the most consistent and readily discernible effect of high rotation rates is to flatten the growth interface. This may be desirable in certain systems where growth from a flat interface has been shown to suppress facet formation which is responsible for radial variations in dopant concentration (*17, 18*). However, in practice, as rotation rates are raised, for example, to 100 rev./min. or above, undesired side effects such as vibration or swing in the crystal mount become increasingly troublesome and may finally act as a cause of serious distrubances in the growth process. Therefore, lower rotation rates tend to be favored. Ruby, sapphire, and YAG, for example, are typically grown at rotation rates in the range 10 to 60 rev./min. These rates are too low to exert any pronounced effect either on interface shape or diffusion layer thickness (7). The main purpose of rotation at all under these circumstances is primarily to even out the effect of any minor thermal asymmetry in the system.

Measurement of temperatures and temperature gradients in such high melting systems as YAG or ruby is difficult to carry out in a meaningful fashion. No satisfactory thermocouples have been found which could yield direct knowledge of subsurface melt temperatures; therefore, one is restricted to surface readings obtained via the use of optical pyrometers or thermopiles. However, an additional complication that further obscures the utility of the readings obtained can be readily inferred by observation of a melt. Quite typically, regardless of the composition of the melt or the type of furnace system used, convection currents will be observed as an ever changing pattern of bright patches outlined by fuzzy dark boundaries. If one utilizes a recording pyrometer on a fixed point of the melt surface, quite wide swings in temperature may be obtained in a matter of seconds [for example, 20° to 30°C. changes have been measured (*8 to 10*)]. The magnitude of the fluctuations and the pattern and rate of change all vary with factors such as furnace geometry and placement of insulation, rate of crystal rotation, crucible rotation, etc. Studies of melt convection processes and factors influencing them have been carried out on lower melting systems and also on model systems at room temperature (7). However, extrapolation of conclusions reached in these studies to behavior of melts in the 2,000°C. temperature range is still fraught with too many uncertainties to afford a quantitative basis for prediction and control purposes. Therefore, an empirical approach must be used in which, for a given system, the behavior of the crystal itself during growth is taken as the best indicator of the thermal environment. Evaluation of many crystals of a number of oxide compositions indicates an apparent correlation between good Czochralski growth conditions and high thermal gradients (but not to the same extent as in the Verneuil process) which are also symmetrical about the center of the melt. Concomitantly the melt convection currents provide useful visual information which can be correlated with changes in thermal environment. A stagnant melt shows no obvious turnover or convection. As longitudinal thermal gradients are introduced, a flow pattern becomes apparent which approximates a spoke wheel, in a symmetrical environment. Further increases in longitudinal thermal gradient result in faster movement and more spokes. Thermal asymmetry causes distortion of the spoke pattern. Practical Czochralski growth experience indicates the desirability of a radially symmetrical thermal environment (*11*); the spoke wheel pattern in the melt is thus one useful hallmark of this condition. Acentric mounting of the crystal with respect to the melt or nonuniform placement of refractory insulation around the crucible result in growth rate variations as the crystal is rotated. This angular variation in growth rate (as the crystal is rotated) in turn encourages normally undesired aberrations such as lopsided growth or marked banding or striations (in the presence of dopants or major impurities).

One problem encountered when Czochralski growth is attempted with a new material is to obtain suitable seeds for the initial runs. This difficulty is typically attacked in one of several different ways. Perhaps the most straightforward method is to dip a platinum or iridium wire or rod into the melt and begin pulling in the usual fashion. Initially a polycrystalline mass will form, but the chances are usually good that one of these crystallites will show preferential growth in the pulling direction and, given sufficient time, will eventually predominate in size over its original neighbors. The orientation of this single crystal can be determined afterwards via x-ray or optical techniques. If it is sufficiently large, it can be cut to yield seeds oriented as desired. Otherwise it can be used as such, as a seed for further growth. Sometimes a number of orientations in the original polycrystalline mass will have nearly equal growth propensities in the direction of pull such that no single crystallite will predominate in a reasonable time. Then one must resort to necking in the pulled mass sufficiently so that hopefully only one of the single crystal components is ultimately in contact with the melt then broadening out again on this seed, whatever its orientation. Yet another approach is to allow a melt to solidify very slowly. Single crystals of sufficient size to use as seeds in subsequent runs may sometimes be mined from the crucible after such a procedure has been followed. Although the internal quality of seeds initially obtained by such techniques may be relatively poor, subsequent growth additions under conditions that are progressively more controlled finally yield satisfactory seed stock.

DISCUSSION AND COMPARISON OF SELECTED OXIDE CRYSTALS PRODUCED BY MELT-GROWTH PROCESSES

The Czochralski process has been used with notable success in recent years to produce laser quality single crystals of ruby and neodymium: YAG in large size ranges never before achieved. Figure 9 illustrates a ruby approximately 2 in. in diameter by about 18 in. long. Figure 10 shows the typical appearance of as grown neodymium: YAG crystals.

The Verneuil process was utilized earlier to grow ruby crystals as long as 14 to 16 in. However, the largest diameter that could be grown routinely was in the range of 1 to 1⅛ in. Similarly, the optical quality and degree of crystallographic perfection of the Czochralski grown ruby surpasses that produced by the Verneuil process. Table 2 presents a comparative summary of representative properties. The most striking differences between the Verneuil and Czochralski grown ruby are the virtual absence of low angle grain boundaries and the much lower dislocation densities in the latter. The optical quality of Czochralski ruby is also correspondingly higher than that of similar size rods fabricated from Verneuil stock, as indicated by typical evaluation test results summarized in Table 2.

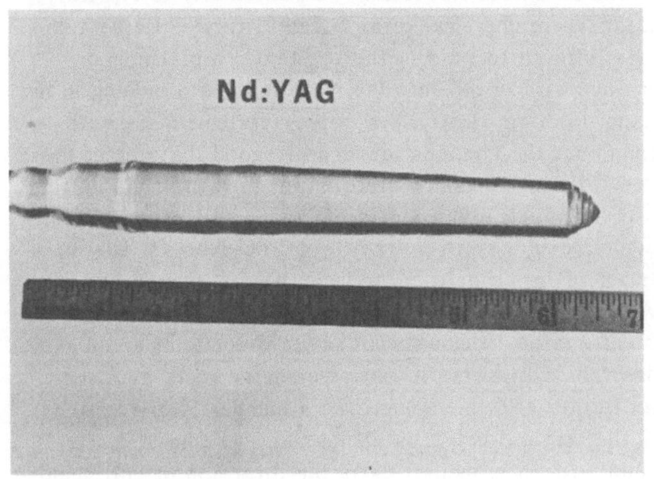

Fig. 9. As grown Czochralski ruby.

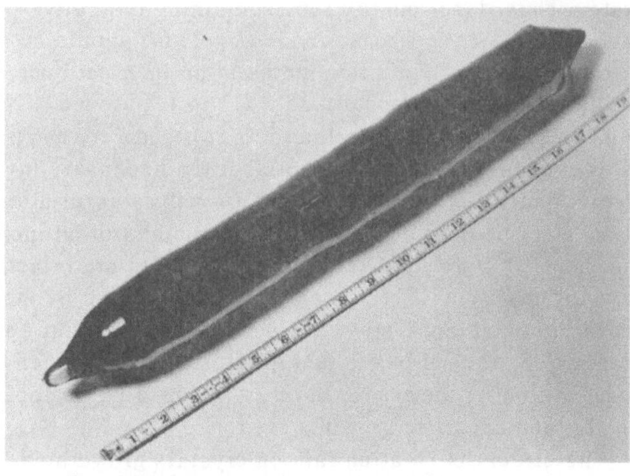

Fig. 10. As grown Czochralski YAG.

It should also be noted here that ruby has been produced in experimental quantities by other approaches such as flux, hydrothermal, and vapor growth methods. However, only the Czochralski process has proved uniquely well suited for routine commercial production of the large sizes, controlled shapes, and reproducibly high optical quality required for laser applications.

The history of the development of neodymium: YAG as a commercial laser crystal bears many similarities to that of ruby, although the time scale is considerably contracted. Lasing action in neodymium: YAG was first obtained in 1964 with a small rod fabricated from a Verneuil grown crystal (*14*), although the same report indicated that better performance was subsequently obtained with flux and Czochralski grown crystals. The Verneuil process proved unsatisfactory, possibly because of the greater structural complexity of YAG as compared with ruby. Flux processes, although capable of yielding high quality crystals, place undesirably restrictive limitations on the size and shape of the crystals grown. The Czochralski process looked immediately attractive when applied to the growth of undoped YAG. However, addition of the desired neodymium oxide dopant to the melt caused several problems which had to be overcome. The most serious of these was that as the neodymium oxide content of the melt was raised, control of the growth process became increasingly troublesome, and crystal quality deteriorated correspondingly rapidly (Nd^{+3} levels in the 1 to 2 atomic % range are required in the YAG host for optimal laser performance). Consideration of the phase diagrams for the systems yttrium oxide-aluminum oxide and neodymium oxide-aluminum oxide (*15*) indicates the existence of a stable garnet phase in the former but not in the latter. Furthermore, in contrast to the situation with ruby wherein aluminum oxide and chromic oxide form a complete series of solid solutions (*16*), evidently neodymium oxide has only a limited solubility in YAG, and marked rejection of Nd^{+3} occurs at the growth interface as measured by the relatively small distribution coefficient.

TABLE 2. COMPARISON OF VERNEUIL AND CZOCHRALSKI RUBY (4, 23)*

	Czochralski	Verneuil
Maximum Crystal Size	2 in. diam. x 14–18 inches long	1 to 1⅛ in. diam. by 14–16 inches long
Strain in as-grown crystals	Low strain - as-grown crystals can be fabricated without prior annealing	Highly strained; must be annealed prior to fabrication
Misorientations	Extremely low. No misorientations detected down to 2 arc-minute limit of the Schulz-Wei X-ray technique. Less than 30 arc-seconds of misorientation measured over 1 cm^2 areas by double crystal X-ray spectrometer method.	High - typical range of 1/4° to 2° over cm^2 regions
Dislocation Densities (in basal plane)	~ 1000/cm^2 average although sizeable regions have been found with 100/cm^2 or lower.	Typically 2-3 orders of magnitude higher than in Czochralski ruby.
Scattering Sites (Defined as voids or inclusions larger than about 1/2 μ causing large angle light scatter measured in %/cm length.)	< .1 %/cm of length	~ Twice the level typical of Czochralski ruby
Performance in Passive Optical Tests		
Twyman-Green Interferometer	<2 fringes per inch length	~twice the fringe count/inch length of equivalent diameter Czochralski ruby
Far Field Pattern	80% of transmitted light flux in 0.2 milli-radians/inch of crystal length	80% of transmitted light flux in 0.4 milli-radians/inch of crystal length

*Typical values for large numbers of 60° oriented rubies (i.e., angle between longitudinal crystal axis and c-axis is 60°).

The effects of this situation were as follows. As growth proceeded, a neodymium rich layer of melt built up just ahead of the growth interface. Then either minor temperature fluctuations occurred which caused deposition of a crystalline layer with such an excess of neodymium oxide that the solubility limit of a neodymium rich phase was exceeded, or constitutional supercooling in effect was responsible for the same end result. The practical solution to this problem was to reduce the growth rate to about one tenth of the value that was used in the growth of the undoped YAG crystals.

A second problem encountered in the growth of neodymium: YAG crystals is the prominent appearance of a core region, a characteristic optical inhomogeneity running longitudinally through the crystals. When neodymium: YAG crystals between crossed polars are viewed through the side, the core appears as a bright band, as shown in Figure 11. If the ends of a crystal are sawed off and polished flat, the core region can also be easily seen in ordinary light (Figure 12). Further study revealed that the shape and spread of the core region were responsive to changes in growth orientation and thermal conditions in the furnace. The tightest cores with the smallest areal extent compared with overall crystal cross-sectional area are obtained with a (111) oriented crystal grown under thermal conditions which favor a very steep interface. Fortunately, the optical distortions due to coring in YAG can thus be confined to the central region of the crystal, and high quality Laser rods can still be fabricated from the surrounding material. Core formation is not restricted to YAG and was earlier observed in other single crystal materials such as silicon and germanium (*17*, *18*); how-

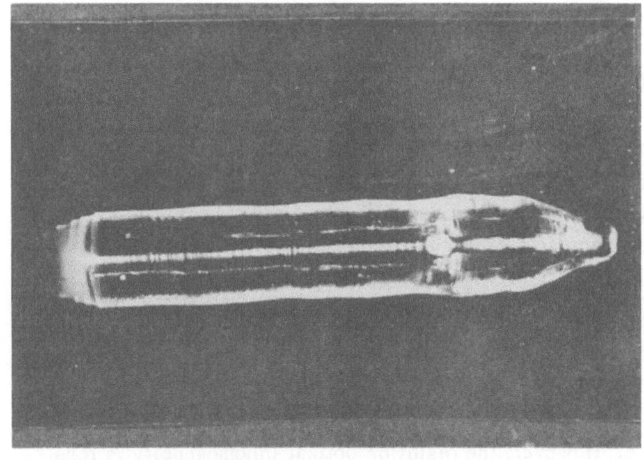

Fig. 11. Appearance of core in neodymium: YAG crystal between crossed polars.

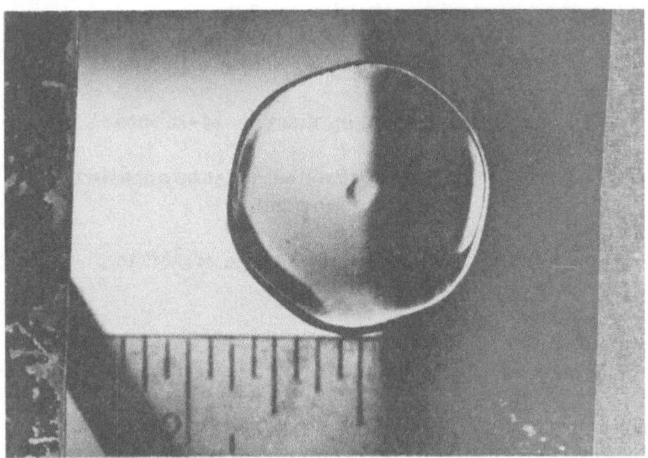

Fig. 12. Cross-sectional view of core in neodymium: YAG crystal, ordinary light. Crystal was back shadowed to increase contrast of core region (~ 4 × mag.).

ever, a common cause seems applicable to every case, namely, formation on the growth interface of facets whose distribution coefficients for melt dopants or impurities may characteristically vary widely from that of the remainder of the growth interface (*17, 18*). This means that the dopant concentration in the facet region will differ from that in the surrounding crystal matrix, a situation which gives rise to internal strain and associated optical inhomogeneities. The facets tend to appear at those regions of the growth interface which are tangent to the crystallographic plane of the facet (211) in the case of YAG.* Depending upon thermal conditions in the furnace, the growth interface of a neodymium: YAG crystal may be gently rounded, a situation which favors broadly spread, large facets, or the interface may be long and steep, a condition which confines the faceting to a very restricted region in the vicinity of the rounded tip end of the crystal.

Although, as noted earlier, melt growth processes are inapplicable to crystal systems which undergo gross structural transformations of a reconstructive nature, displacive transformations tend to be significantly less disruptive (in the sense that large density changes do not occur) and may be tolerated provided that other side effects such as twinning can be eliminated. Examples of materials displaying such behavior are $LaAlO_3$, which was investigated earlier as a possible laser host for Cr^{+3} or Nd^{+3} (*19*), and more recently $Ba_2NaNb_5O_{15}$, crystals of which have performed outstandingly as an electrooptic frequency doubler for 1.06 μ laser radiation (*20*). $LaAlO_3$ undergoes a change from cubic to rhombohedral as the crystal is cooled below the transition temperature of 512°C., and this structural shift causes extensive twinning. In $Ba_2NaNb_5O_{15}$, twinning is associated with a change from tetragonal to orthorhombic as the crystal is cooled below 300°C. In both cases, the extensive formation of twin boundaries throughout the structure is responsible for such an increase in internal light scattering that the crystals in the as grown condition are opaque at room temperature. Fortunately, a simple method has been found for eliminating these boundaries; application of mechanical stress along certain preferred crystallographic directions can convert a specimen into a single twin of quite acceptable optical properties. With $LaAlO_3$, application of relatively small loadings of ~ 2,000 lb./sq. in. along the (111) axis of fabricated rods or bars at room temperature suffices. In the case of $Ba_2NaNb_5O_{15}$, the crystal is heated back through the transition temperature (at which point the twinning spontaneously disappears), and sufficient stress is applied along an a or b axis to prevent retwinning as the specimen is cooled again to room temperature (*20*).

CURRENT TRENDS IN THE MELT GROWTH OF OXIDE CRYSTALS

LARGER SIZES

Experience of the recent past clearly indicates the strength of the demand for ever larger size laser crystals. The specimens of Czochralski grown ruby and neodymium: YAG shown in Figures 9 and 10 illustrate the present day capabilities of the growth systems developed to meet this demand. These size levels are probably adequate for most of the currently anticipated applications. If necessary, even larger crystals can undoubtedly be produced by appropriate scale-up of present facilities. However, some ultimate limitations on size must eventually be encountered with the Czochralski process as is shown by the following considerations. Larger crystals require a larger melt reservoir which must in turn be contained in ever more voluminous crucibles. As the diameter of a crucible is increased, the wall temperature required to maintain the material near the top center surface at the melting point also increases (that is, the average radial thermal gradient across the melt surface must increase). The obvious limit to this trend is set by the melting point of the crucible. If crystal sizes are required that exceed this limit, some alternative method of containing large melts must be developed (by assuming the unavailability of a still higher melting crucible material that would also be compatible with the melt). Skull melting, wherein a melt is contained in a solidified shell of identical composition, may someday prove to be an acceptable solution to the problem. However, skull melting technology as applicable to the growth of laser grade ruby or neodymium: YAG is still in an embryonic stage of development, and severe engineering problems remain to be

*Core formation is also observed in Czochralski grown ruby as a result of formation of r plane facets on the growth interface. However, the resulting optical inhomogeneity is less pronounced than in the case of YAG and can best be observed by schlieren techniques on crystals with polished ends (*4*).

solved. Alternatively, the need for maintaining a large reservoir of melt throughout a run could be dispensed with altogether if a suitable technique were developed for float zone growth of very large diameter crystals. Present diameter limits for this process are around ½ in.

SPECIAL SHAPES

Emphasis to date has been most heavily focused on utilizing the Czochralski process to produce massive crystals with essentially circular symmetry which are best suited for subsequent fabrication into laser rods. However, increasing interest is evidenced in other shapes not necessarily intended for laser use, for example, sapphire slabs for special optical applications, sapphire fibers for use as metal reinforcement, sapphire tubes for high intensity radiation lamp envelopes, etc. The Verneuil process has, of course, long been utilized for the production of sapphire disks and domes, but the other shapes have thus far been beyond the scope of Verneuil technology. Very recently, however, important progress has been made in learning how to adapt the Czochralski process to the growth of a variety of shapes such as those listed above. In all cases success depends upon careful shaping and control of thermal gradients in the vicinity of the growth interface. In order to attain the necessary thermal conditions, drastic departures from customary Czochralski furnace geometry may be required. Thus, one recent report discusses the use of a floating orifice to produce small diameter sapphire shapes such as fibers (*21*). It is anticipated that further developments along these lines will ultimately permit melt growth of an even broader variety of shapes, not only of sapphire but other oxide single crystals as well.*

ACKNOWLEDGMENT

The author wishes to thank F. R. Charvat, O. H. Nestor, and J. C. Smith for their helpful comments and constructive criticisms during the course of preparation of this paper. All crystals shown herein were prepared by the Crystal Products Department of Union Carbide Corporation.

LITERATURE CITED

1. General References on Crystal Growth Theory, Techniques, and Experimental Methods.
   a. Buckley, H. E., "Crystal Growth," Wiley, New York (1951).
   b. Gilman, J. J., ed., "The Art and Science of Growing Crystals," Wiley, New York (1963).
   c. Lawson, W. D., and S. Nielsen, "Preparation of Single Crystals," Academic Press, New York (1958).
   d. Peiser, H. S., ed., "Crystal Growth; Proceedings of an International Conference on Crystal Growth, Boston, 20–24 June 1966," Pergamon Press, New York (1967).
2. General References Which Emphasize Theoretical Aspects of Melt Growth.
   a. Pfann, W. G., "Zone Melting," 2 ed., Wiley, New York (1966).
   b. Chalmers, B., "Principles of Solidification," Wiley, New York.
   c. Zief, M., and W. R. Wilcox, ed., "Fractional Solidification," M. Dekker, New York (1967).
3. Maiman, T. H., *Nature*, **187**, 493–494 (1960).
4. Charvat, F. R., J. C. Smith, and O. H. Nestor, "Crystal Growth; Proceedings of an International Conference on Crystal Growth,' p. 45, Pergamon Press, New York (1967).
5. Kestigan, M., and W. W. Holloway, Jr., *ibid.*, p. 453.
6. Cockayne, B., *Phil. Mag.*, **12**, No. 119, 943–50 (1965).
7. Robertson, D. S., *Brit. J. Appl. Phys.*, **17**, 1047–1050 (1966).
8. Wilcox, W. R., and L. D. Fullmer, *J. Appl. Phys.*, **36**, No. 7, 2201–2206 (1965).
9. Brice, J. C., and P. A. C. Whiffin, *Brit. J. Appl. Phys.*, **18**, 581–585 (1967).
10. Cockayne, B., and M. P. Gates, *J. Material Sci.*, **2**, 118–123 (1967).
11. Morizane, K., A. Witt, and H. C. Gatos, *J. Electrochem. Soc.: Solid State Sci.*, **114**, No. 7, 738–742 (1967).
12. Burton, J. A., R. C. Prim, and W. P. Slichter, *J. Chem. Phys.*, **21**, No. 11, 1987–1991 (1953).
13. Burton, J. A., E. D. Kolb, W. P. Slichter, and J. D. Struthers, *ibid.*, 1991–1996.
14. Geusic, J. E., H. M. Marcos, and L. G. Van Uitert, *Appl. Phys. Lett.*, **4**, No. 10, 182–184 (1964).
15. Levin, E. M., C. R. Robbins, and H. F. McMurdie, "Phase Diagrams for Ceramists," Figures 311 and 312, The American Chemical Society, Inc., New York (1964).
16. *Ibid.*, Fig. 309.
17. Dikhoff, J. A. M., *Solid-State Electronics*, **1**, 202–210 (1960).
18. ———, *Philips Tech. Rev.*, **25**, No. 8, 195–206 (1963/64).
19. Fay, H., and C. D. Brandle, "Crystal Growth; Proceedings of an International Conference on Crystal Growth," p. 51, Pergamon Press, New York (1967).
20. Geusic, J. E., H. J. Levinstein, J. J. Rubin, S. Singh, and L. G. Van Uitert, *Appl. Phys. Lett.*, **11**, No. 9, 269 (1967).
21. LaBelle, H., Jr., and A. I. Mlavsky, *Tech. Rept. AFML-TR-66-246* (Aug., 1966).
22. Faust, J. W., Jr., "Crystal Growth; Proceedings of an International Conference on Crystal Growth," pp. 183–185, Pergamon Press, New York (1967).
23. Plooster, M. N., O. H. Nestor, J. C. Smith, and R. L. Hutcheson, paper presented at Annual Meeting of American Ceramic Society (May 4, 1965).
24. Alford, W. G., and W. H. Bauer, "Crystal Growth; Proceedings of an International Conference on Crystal Growth," p. 71, Pergamon Press, New York (1967).

---

*Germanium and silicon have been grown as webs (that is, micron thickness ribbons 1 to 2 cm. wide) via a modified Czochralski process for some years now; however, the specific technique, involving rapid growth from a supercooled melt via a twin plane reentrant edge mechanism, has not yet been successfully applied to oxide systems (*22*).

THE PROGRESSIVE MODE OF MULTISTAGE CRYSTALLIZATION

G. R. Atwood

The following is a comparative analysis of two types of continuous multistage crystallization schemes. The first represents a reflux process in the classical sense, whereas the second is analogous to zone refining (and includes continuous zone refining as an example). The latter scheme, designated the *progressive mode* because the charge essentially progresses along an infinite series of stages, is shown to be up to twice as efficient as the classical reflux mode.

Continuous multistage separation schemes classically incorporate a reflux to take advantage of the multiplicity of stages available. Without this reflux, the total possible separation under steady state conditions is no better than that which could be obtained with a single stage. The maximum possible separation is obtained at total reflux, in which case each stage is able to perform a full stage separation. If the separation factor per stage is expressed by s, then the total separation factor S for N stages is given by s^N. However, at total reflux there can be no product, and practical multistage operations are carried out at some partial reflux which, in conjunction with the appropriate number of stages, can be optimized for the specific separation required.

With the invention in 1952 of zone melting by Pfann (*1*), there was automatically introduced into our technology a new scheme of multistage operation, one which does not incorporate reflux in the classical sense. The technique of zone refining has had its major impact as a batch process, but it can be made continuous as described by Pfann, Kennedy, and others (*2*). The schematics of the process are independent of whether it is carried out in ingot form or otherwise. If one considers the zone itself as the separation stage, an analogy which will be more obvious when we discuss rotary drum separations, one can consider the number of stages in a zone refining operation as being equivalent to the number of zones which are passed simultaneously through the ingot. Like a reflux process, feed can be introduced into the ingot, and products can be removed from both ends. We have chosen to designate this mode of operation as the *progressive mode*, not implying modern or advanced, but defining the fact that the stages progress through the process charge, or, as in some possible systems, the process charge progresses through an infinite series of stages.

Although there is no reflux in the progressive mode, there is a variable rate of product removal (relative to a unit rate of zone passage or crystal growth). When this rate is zero, thus comparable to the total reflux type of operation, the maximum possible separation is achieved. This is of the order of, or better than, $S = s^M$. This ultimate separation is superior to that for the reflux type of operation in that for comparable solid—liquid geometry in both cases, N will of necessity be less than M. Practically, continuous zone melting will operate at a finite product removal rate, and that rate, in conjunction with the appropriate ingot length, can be optimized for the production of a specified product. It is the purpose of this work to compare the optima of the reflux and progressive modes.

COMPARATIVE SCHEMATICS OF THE TWO PROCESSES

Figure 1 shows the comparative schematics for the reflux and progressive modes. In both cases, we will imagine long ingots with a number of equispaced molten zones within them. The heaters for these zones are

Union Carbide Corporation, Tarrytown, New York.

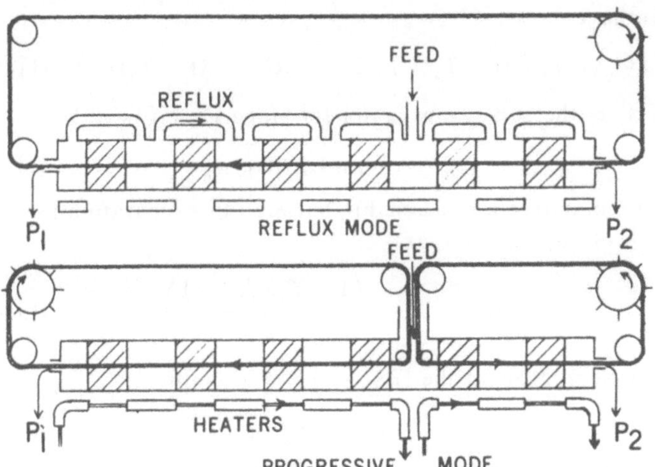

Fig. 1. Comparative schematics for the reflux and progressive modes of multistage melt crystallization.

shown below the ingot containers. The heaters for the reflux mode are stationary relative to the container, but for the progressive mode, they move as indicated by the arrow. Both schemes provide for feed ports along with takeoff ports for the purified P_1 and stripped P_2 products.

The ingot for the reflux mode is pulled within its container toward the P_1 end at some practical rate which we can define as unity. Although the ingot as a whole moves, the positions of the solid and liquid zones do not change, the solid zone melting on its advancing face and crystallizing on its receding face at the same unit rate as the ingot is moved. If there were no reflux, product P_1 would be obtained also at unit rate. The tubes shown provide for reflux, and the takeoff rate T_1 is given by

$$T_1 = 1 - R \tag{1}$$

The takeoff rate T_2 of the crude product P_2 is

$$T_2 = F + R - 1 = F - T_1 \tag{2}$$

There is no reflux for the progressive mode, and in order to obtain product P_1 at a rate T_1, the purification section of the ingot must be pulled within its container toward P_1 at the rate T_1. For a unit crystal growth rate, therefore, the heaters below the purification section of the ingot must move along the container toward P_2 at a rate of $(1 - T_1)$. Since product P_2 is obtained at a rate T_2, the ingot at the stripping end of the column must be pulled at a rate T_2, and the rate of heater motion at this end must be $(1 + T_2)$ or $(1 + F - T_1)$. An alternative to having different rates of heater motion along the two sections of the column would be to let the heaters under the stripping section also move at $(1 - T_1)$, in which case the crystal growth rate in this section would be $(1 - F)$.

In both schemes, the separation obtained is, of course, the result of the preferential distribution of the solute or impurity in the liquid rather than in the growing crystal phase. The distribution coefficient k is defined as the ratio of the impurity in the solid to that in the liquid from which the crystal is growing, and we will limit our analyses to only those circumstances in which k is less than unity. Although the distribution coefficient is in practice a complex function of concentration and crystallization rate, we will presume k for our purposes to be constant for a given chemical system.

DEFINITION OF EFFICIENCY

The cost of operating a multistage process should be proportional to the number of stages involved, provided that the same complete phase transitions occur at each stage and that thermal losses and other inefficiencies related to stage operation are identical. This would not be true for a process such as distillation, in which all stages but one involve only interphase contact. It should be true for the schematic processes in Figure 1. The costs will also be inversely related to the rate at which feed is processed, so that the efficiency E of the process can be defined by

$$E = F/N \tag{3}$$

that is, the quantity of feed processed per unit of crystal grown. The number of stages necessary to separate the feed into the specified products, P_1 and P_2, will be related to the rate of feed consumption, being a minimum when $F = 0$ and infinite when F is some maximum possible value (corresponding to the minimum reflux ratio). At intermediate values of F and the comparable N, the F/N ratio can be maximized.

The maximum possible efficiency is, of course, limited by the difficulty of the separation. For the purification of a compound, we can define a separation factor S_1 for the purification section of the column as the ratio of impurity (solute) in the feed to that in the product P_1. Similarly, there will be a separation factor S_2 for the stripping section, defined as the ratio of solute in product P_2 to that in the feed. The separation being performed is defined by both S_1 and S_2, and the ratio of products P_1 and P_2, that is, T_1/T_2, is given by

$$\frac{T_1}{T_2} = \frac{S_1(S_2 - 1)}{(S_1 - 1)} \tag{4}$$

and the efficiency for the specific separation can be expressed by

$$E = T_1\left(1 + \frac{(S_1 - 1)}{S_1(S_2 - 1)}\right)\Big/N \tag{5}$$

We can therefore express the efficiency in terms of the takeoff rate of P_1 alone and can, for the specific job, redefine a relative efficiency E' as

$$E' = T_1/N \tag{6}$$

that is, the quantity of good product produced per unit of crystal grown. For large values of S_1 and/or S_2, E' is almost equivalent to E, inasmuch as T_1 is almost equal to F.

It would be feasible to determine optimum efficiencies over a wide range of values of S_1 in conjunction with a wide range of values of S_2 and for a range of k values. Practically, this would be laborious even with the aid of computer facilities owing to the large number of combinations involved. It is possible, however, to determine the number of stages N_1 for the purification section of the process as a function of S_1 and T_1 and define an efficiency E_1 for that section alone as

$$E_1 = T_1/N_1 \tag{7}$$

It is not similarly possible to compute the number of stripping stages N_2 as a function of S_2 and T_2 alone, the operation of the stripping section being dependent upon that of the purification section. The analysis is therefore greatly simplified if we restrict it to the purification section only. This approach is justifiable in view of the fact that zone processes (and their reflux counterparts) would be used predominantly for ultrapurification, in which case high values of S_1 and N_1 are required. Small values of S_2, however, are usually sufficient to reduce the product P_2 to such a crude grade that further concentration of solute by melt crystallization is impractical, and an accessory technique must be used to strip the desired compound for recycle as new feed. In the theoretical analyses that follow, therefore, only the efficiencies of the purification sections will be compared.

ANALYSIS OF THE REFLUX MODE

The analysis of the reflux mode follows classical steady state solute balance techniques. The quantity of solute entering a given liquid zone i is the sum of that refluxed from liquid zone $(i-1)$ and that melted out of the solid zone $(i+1)$, see Figure 2. Letting C_S and C_L be the solute concentrations in the solid and liquid, respectively, we have

$$\text{solute entering } i = C_{S(i+1)} + (1-T_1)C_{L(i-1)} \tag{8}$$

The solute leaving is that refluxed from liquid zone i and that frozen into the growing solid zone i:

$$\text{solute leaving } i = C_{Si} + (1-T_1)C_{Li} \tag{9}$$

These must be equal at steady state, and if

$$k = C_{Si}/C_{Li} \tag{10}$$

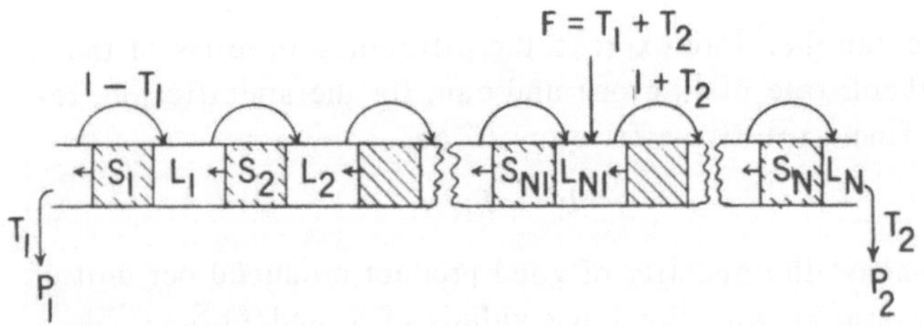

Fig. 2. Schematic diagram for multistage crystallization with reflux.

we have

$$kC_{L(i+1)} + (1-T_1)C_{L(i-1)} = kC_{Li} + (1-T_1)C_{Li} \tag{11}$$

If s_i is the separation factor for stage i, that is

$$s_i = C_{Li}/C_{L(i-1)} \tag{12}$$

we can derive a relationship between the separation factors of adjacent stages:

$$s_{(i+1)} = 1 + \left(\frac{1-T_1}{k}\right)\left(\frac{s_i - 1}{s_i}\right) \tag{13}$$

The separation at the first stage is given by

$$s_1 = 1/k \tag{14}$$

and it is thus possible to compute the separation factor for any other stage by

$$s_i = \frac{1 + [1 + Q + Q^2 \ldots + Q^{(i-1)}]\,(1-k)}{s_{(i-1)}\,s_{(i-2)}\,s_{(i-3)} \ldots s_2} \tag{15}$$

where

$$Q = (1-T_1)/k \tag{16}$$

The overall separation factor for the purification section S_1 is given by

$$S_1 = s_1 s_2 s_3 \ldots s_{N_1} = \prod_1^{N_1} s_i \tag{17}$$

and from Equation (15) we get

$$\prod_1^{N_1} s_i = 1 + \frac{(1-k)}{k}\sum_1^{N_1} Q^{(i-1)} \tag{18}$$

It can be demonstrated that

$$\sum_1^{n} x^n = \frac{[x^{(n+1)} - 1]}{(x-1)} \tag{19}$$

so that

$$S_1 = 1 + \frac{(Q^{N_1} - 1)(1-k)}{(Q-1)k} \tag{20}$$

We can solve for the necessary number of stages N_1 in the purification section by

$$N_1 = \frac{\log\left((Q-1)(S_1-1)\dfrac{k}{(1-k)} + 1\right)}{\log Q} \tag{21}$$

It can be similarly shown that the necessary number of stages in the stripping section N_2 is given by

$$N_2 = N - N_1 = \frac{\log\left(1 + \dfrac{(W-1)}{Q^{N_1}}\,S_1\,(S_2-1)\,\dfrac{k}{(1-k)}\right)}{\log W} \tag{22}$$

where

$$W = (1 + T_2)/k \tag{23}$$

and T_2 and T_1 are related by Equation (4). We will not, however, make any computations for the stripping section.

As expected, for any given values of k and S_1, there are minimum possible values for the reflux and for the number of stages, and a hyperbolic type of plot of N_1 vs. the reflux ratio $(1 - T_1)/T_1$ can be made, see Figure 3. Of more significance is the plot of efficiency E_1 vs. the number of stages, Figure 4, or the takeoff rate, Figure 5. The maximum value of these curves is the optimum efficiency. Values were computed for several combinations of S_1 and k, and these are the solid lines shown in Figure 6.

ANALYSIS OF THE PROGRESSIVE MODE

An exact algebraic analysis of continuous zone melting is not possible. Nevertheless, relatively good approximations can be made for those separations involving a large number of stages (or a high ingot length to zone length ratio). The analysis given here follows very closely that presented by Pfann (3) but differs in certain details because our schematic model differs slightly from Pfann's zone void technique.

Figure 7 is a schematic concentration profile for the solute in an ingot being refined. Whereas it is our intent to pass many zones along the ingot at one time, the analysis can be made in terms of the sequential passage of single zones. For an ingot of length $L = L_1 + L_2$ having a solute profile as shown by the solid line, it is intended to add a quantity $\theta + \phi$ of feed, concentration C_F,

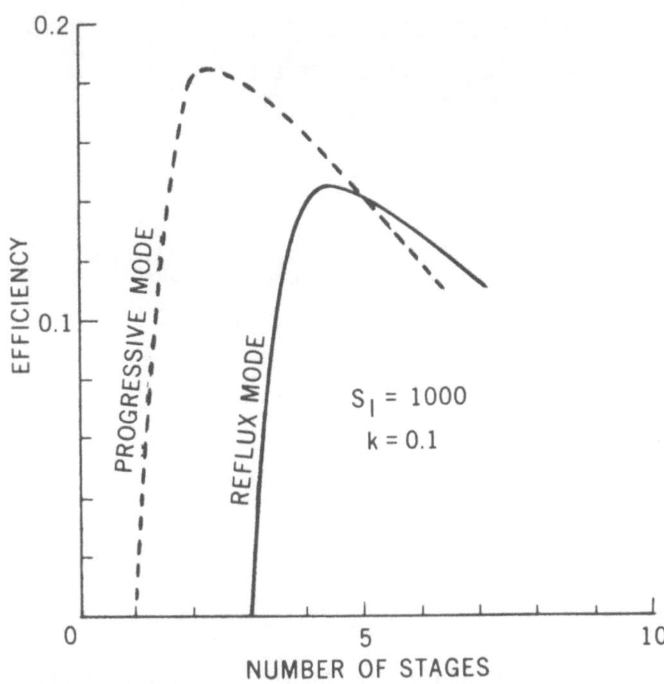

Fig. 4. Purification efficiency vs. the number of stages for the reflux (solid line) and progressive (dashed line) modes for a system with a distribution coefficient of 0.1 and a required purification factor of 1,000. (One stage = two zone lengths.)

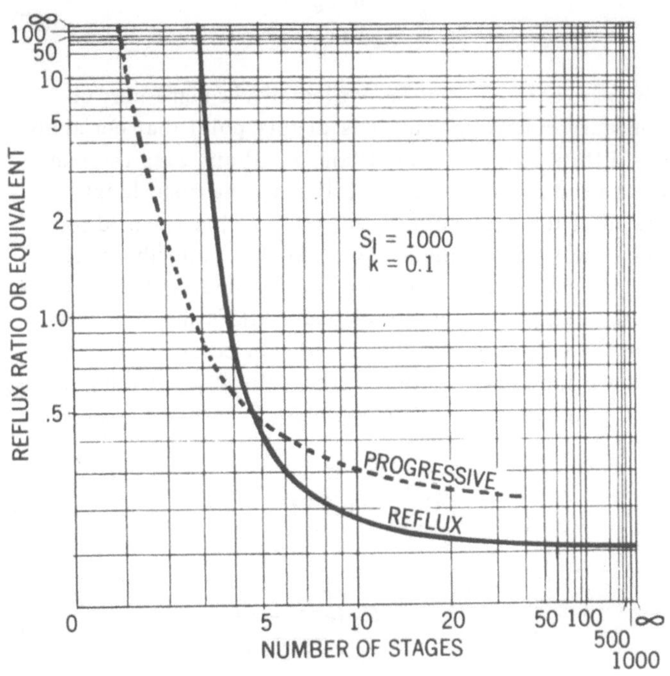

Fig. 3. Plots of reflux ratio (or equivalent) vs. the number of stages in the purification sections for the reflux (solid line) and progressive (dashed line) modes for a system with a distribution coefficient of 0.1 and a required purification factor of 1,000. (One stage = two zone lengths.)

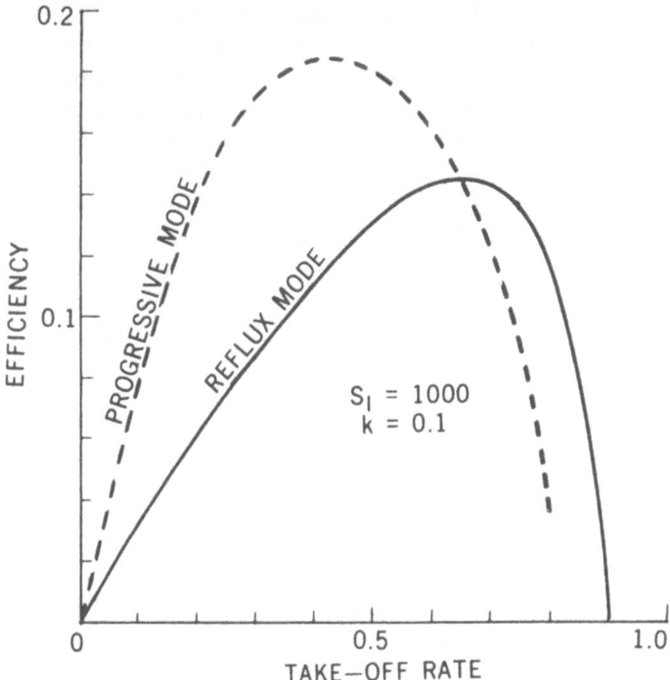

Fig. 5. Purification efficiency vs. the takeoff rate for the reflux (solid line) and progressive (dashed line) modes for a system with a distribution coefficient of 0.1 and a required purification factor of 1,000.

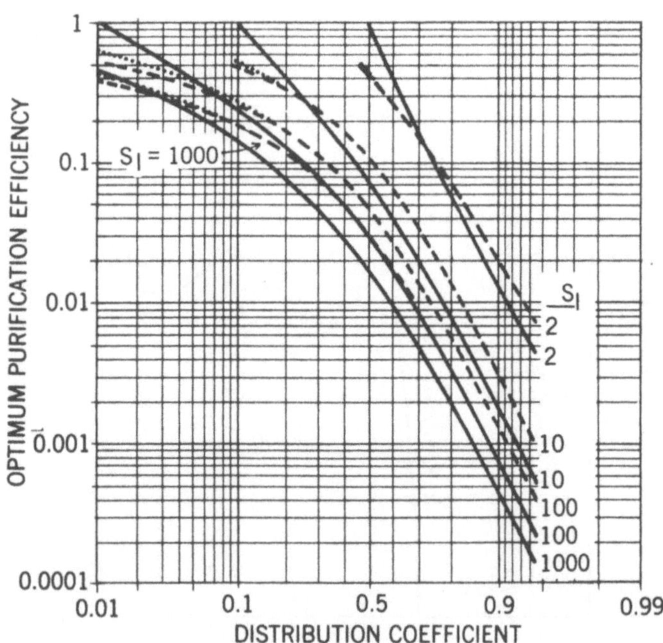

Fig. 6. Optimum purification efficiencies vs. the distribution coefficient for several purification factors (S_1). Reflux mode—solid lines; progressive mode—dashed lines; arithmetically simulated progressive mode—dotted lines. These results are for the purification section of the unit and do not include any computations for the stripping section.

where indicated so as to shift the ingot profiles as shown by the dashed curves. The crosshatched portions of the ingot can then be removed as products P_1 and P_2. The passage of a single zone of length ℓ through the modified ingot should then restore the ingot to its original state. It cannot really do so of course, since the modified profile contains a discontinuity in its slope, which discontinuity must be propagated by successive zone passes toward the P_1 end as discontinuities in the successive derivatives of the slope of the profile. An approximation for the purification section of the ingot can be obtained, however, by overlooking the discontinuity and by presuming the modified profile to follow the dotted curve. The error introduced thereby is not as serious as might be anticipated, particularly for large values of L_1. This was investigated by arithmetical simulation of the zone melting scheme for the smaller values of L_1.

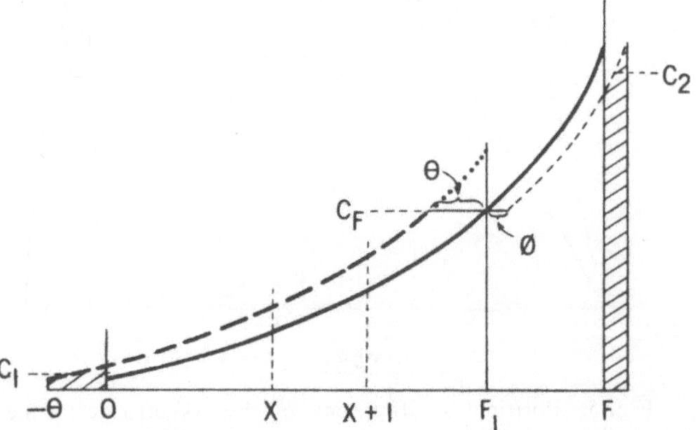

Fig. 7. Solute profile for analysis of continuous zone refining or progressive mode.

Since there is no steady state, but rather a reproducible state, it is necessary to express the solute profile C_x for the ingot in terms of that which would be derived by the passage of a zone of length ℓ through the modified ingot with a solute profile of $C_{(x+\theta)}$:

$$C_x = \frac{k}{\ell}\left[\int_0^{x+\ell} C_{(x+\theta)}\,dx - \int_0^{x} C_x\,dx\right] \tag{24}$$

The first integral represents all that solute dissolved into the zone as it moves from zero to $(x + \ell)$. The second represents all that solute which is frozen out of the zone during the same movement. The difference is that left in the zone, and divided by ℓ it is the concentration of solute in the zone. The concentration C_x in the solid crystallizing from the zone, which latter lies between x and $(x + \ell)$, is the product of k and the concentration in the zone.

The solution to Equation (24) is given by

$$C_x = A\,(e^{Bx} + Z) \tag{25}$$

where

$$Z = \frac{k\,(e^{B\theta} - 1)}{(1 - k)\ell B} \tag{26}$$

and B is related to k by

$$k = \frac{B\ell\, e^{-B\theta}}{e^{B\ell} - e^{-B\theta}} \tag{27}$$

The parameter A must be evaluated via some sort of boundary condition, and it is at this point that our analysis differs the most from Pfann's. Pfann sets the concentration of the solid crystallizing one zone length ahead of the feed point as kC_F (4). This is valid for his zone void technique, since at that point the zone breaks into a large volume of liquid feed, and, with complete mixing, the zone itself must attain the concentration C_F. For our model this will not be so, and we will assume the concentration of the solid crystallizing at the feed point to be C_F. Letting $C_x = C_F$ when $x = L_1$, we have

$$C_F = A\,(e^{BL_1} - Z) \tag{28}$$

which leads to

$$C_x = \frac{C_F(e^{Bx} - Z)}{(e^{BL_1} - Z)} \tag{29}$$

Now the purified product P_1 is not the result of a continuous takeoff of product of composition C_1 but is rather the sporadic takeoff of a variable composition which must average out to C_1. The average product composition is given by

$$C_1 = \frac{1}{\theta}\int_0^{\theta} C_x \, dx \tag{30}$$

which, when integrated, gives

$$C_1 = \frac{C_F}{(e^{BL_1} - Z)}\left(\frac{e^{-B\theta} - 1}{B\theta} - Z\right) \tag{31}$$

To find the necessary ingot length, we must solve for L_1:

$$L_1 = \frac{2.3}{B}\log_{10}\left[S_1\left(\frac{e^{-B\theta} - 1}{B\theta} - Z\right) + Z\right] \tag{32}$$

where S_1 is the separation factor C_F/C_1.

Now Equation (32) was derived on the basis of passing zones individually through the solid ingot of length $L_1 + L_2$. To be analogous to the scheme in Figure 1, it is necessary to accumulate the zones within the ingot. We shall therefore pass N_1 zones into, but not through, the purification section. As each zone is added, a portion θ of product must be removed, and the ingot with its accumulated zones is shortened. The actual physical length L_1' of the purification section is

$$L_1' = L_1 - N_1\theta \tag{33}$$

The quantity of ingot required for a zone is the sum of the zone length and that solid length necessary between zones, that is, $\ell + \ell_S$, so that the number of zones or stages possible is

$$N_1 = \frac{L_1'}{\ell + \ell_S} = \frac{L_1}{\ell + \ell_S + \theta} \tag{34}$$

The passage of a zone is accomplished by the movement of all heaters an interzone length, that is, $\ell + \ell_S$, to the right, and for each passage we remove θ product, that is, the ingot moves θ to the left. The crystal growth rate is thus $\ell + \ell_S + \theta$ for each θ of product. The takeoff rate T_1 must be

$$T_1 = \frac{\theta}{\ell + \ell_S + \theta} \tag{35}$$

and the efficiency is

$$E_1 = T_1/N_1 = \theta/L_1 \tag{36}$$

It will be noted that the efficiency for the progressive mode is independent of the solid-liquid ratio just as it is for the reflux mode.

Similar derivations were made for the stripping section, leading to

$$L_2 = \frac{2.3}{\beta\ell}\log_{10}\frac{\beta\phi}{(1 - e^{-\phi})}[S_2(1 + Y) - Y] \tag{37}$$

where β is related to k and ϕ by

$$k = \frac{\beta\ell e^{-\phi}}{e^{\beta\ell} - e^{-\beta\phi}} \tag{38}$$

and

$$Y = \frac{k[S_1(\theta + \phi) - \theta]}{\ell S_1(1 - k) - k\theta(S_1 - 1)} \tag{39}$$

This derivation ignores the break in the solute profile as the last zone length freezes and hence is, like that for the purification section, approximate rather than exact. It will be noted that the parameters for the stripping section depend, as in the reflux mode, upon the degree of separation S_1 in the purification section. The efficiency E_2 for the stripping section depends upon whether the zones in the two sections are passed at equal linear rates or so as to give equal crystal growth rates. Derivations of the efficiencies for these two cases were not made.

Computations were made for the purification section by first evaluating k by Equation (27) for combinations of B and θ. Plots were then constructed for the extraction of B values for k, θ combinations, and values of L_1 were computed for k, θ, S_1 combinations. For defined values of k and S_1, a plot of $(1 - T_1)/T_1$, the equivalent to a reflux ratio, vs. N_1 is similar to that for the reflux mode. Such a plot is shown in Figure 3, dashed line, for the situation where $\ell = \ell_S$, and thus $N_1 = L_1/2$. Figures 4 and 5 show how plots of the efficiency E_1 vs. the number of stages and the takeoff rate, respectively, compare with those for the reflux mode. It will be noted that there are some operating conditions under which the progressive mode is superior and others under which the reflux mode is superior. Of significance, however, is the fact that when each process is operated at its own optimum (different values of N_1 and T_1 for each mode), the progressive mode is the most efficient.

Plots of optimum efficiencies vs. k for various values of S_1 are shown as the dashed lines on Figure 6. The dotted lines on this plot are the results of arithmetical simulations of the progressive mode for the smaller values of L_1 and show how the derived algebraic relationships break down in this range. A comparison of the two modes shows that the progressive mode is superior by a factor of up to 2 for difficult separations, that is, high values of k and S_1. The reflux mode is superior only for very easy separations requiring less than two stages. Inasmuch as the curves plotted in Figure 6 are for the purification sections alone, it was considered possible that the inclusion of values for the stripping sections might substantially change the relationships for the easier separations.

ARITHMETICAL SIMULATIONS

It was felt that the approximations involved in the development of the Equations (37), (38), and (39) for the stripping section of the progressive mode prohibited any valid computations thereby. Therefore, computerized arithmetical simulation of the process was used to obtain data for the unit as a whole (stripping plus purification). For this purpose an ingot of 1,000 segments was used, and zone passage was accomplished by alternately melting one segment into and freezing another segment out of the zone, the latter process being carried out so that the ratio of the impurity level in the frozen segment to that in the remaining zone was k. Feeding was accomplished after each zone pass by splitting the ingot at the position of unit concentration level and shifting the sections T_1 and T_2 segments, respectively, the voided segments being reset to unit concentration. T_1 and T_2 segments were removed from the ends as products. The efficiency was defined as in Equation (3) as the feed $(T_1 + T_2)$ divided by the ingot length (1,000), that is, the quantity of crystal which had to be grown. For a given zone length, zone passes were alternated with feeding until a steady state of constant product purity was reached. Trials were made over a range of zone lengths to find that optimum length at which the best product could be produced for a predefined efficiency $(T_1 + T_2)/1{,}000$ and recovery $T_1/(T_1 + T_2)$.

Optimum purification factors for 90% recovery are plotted in Figure 8 for several values of k and E. The curves are hyperbolic in form and were replotted against the reciprocal of the efficiency, Figure 9, where it will be observed that they become quite linear. Data for the optimized reflux mode are also shown as obtained from Equations (21) and (22).

Plotting the data from Figure 9 against k as in Figure 6 gives Figure 10. Here it is observed that the factor of 2 by which the efficiency of the progressive mode is superior to that of the reflux mode is applicable over the entire range of separation difficulty, that is, low values of k as well as high values. Data have not been computed for recovery values other than 90%, but there is no reason to expect substantial differences. Computations at other recovery levels are now in progress.

DISCUSSION

The results shown in Figures 6 and 10 are, of course, a function of the basic assumptions involved in their development, that is, a constant value of k, regardless of concentration or crystallization rate, and complete mixing in the liquid zones. To the extent that these ideal assumptions are not correct, the efficiencies calculated would not be valid. However, the deviations from ideality would be comparable for both schemes of operation, and the relative ratios of the efficiencies calculated should hold. It is difficult to conceive any condition of operation which would lead to a change in the relative superiority of the two schemes, lest perhaps it be the antiquity of the crystal being grown. Whereas in the progressive mode a growing crystal progresses through

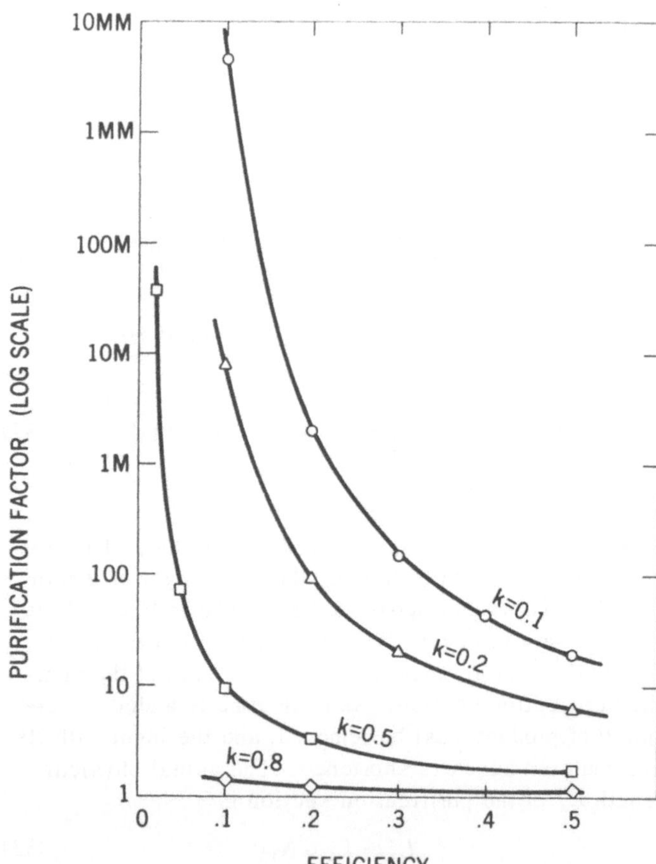

Fig. 8. Arithmetically simulated continuous zone refining or progressive mode; optimum purification factors for operation of the complete unit (purification plus stripping) at several predefined efficiencies and 90% recovery.

the entire ingot and is then terminated, the crystals in the reflux mode are perpetuated indefinitely. Advantages or disadvantages resulting thereby are dependent upon the complexities of crystal ripening or degradation and the relative tendencies to initiate and perpetuate defects, a subject beyond the scope of this comparison.

It is important to note that the efficiencies shown in the figures are independent of the solid-liquid ratio for both schemes of operation. This means that the relative efficiencies of the two schemes will hold regardless of the type of crystallizer used. So far we have discussed only the analogy to zone refining per se, but more conventional crystallization techniques can be used. Some of these will be more easily operated in one mode than another, but in those cases where both schemes can be applied, our comparisons will be valid. Pfann discusses the rotating drum crystallizer (5) but does not discuss the details of reflux or, alternatively, stage progression necessary to make the process continuous. Figure 11 shows how the system can be used in either scheme. Operation in the reflux mode is obvious. In the progressive mode there would be no reflux, and the drums, along with baffles and agitators, would move toward the right along the trough of liquid. Drums would be re-

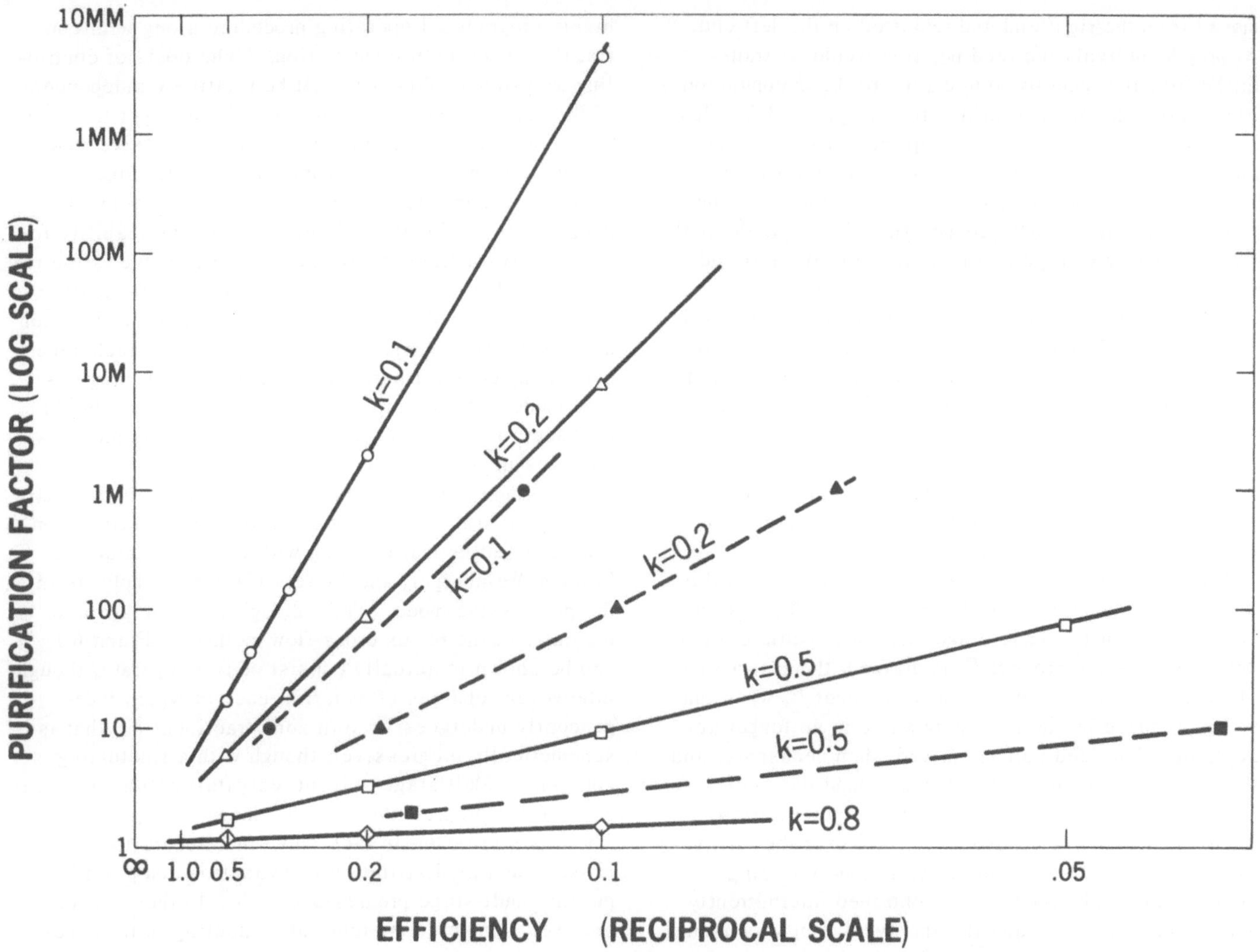

Fig. 9. The results of Figure 8 as plotted on a reciprocal efficiency scale. Dashed lines are for the reflux mode as computed via Equations (21) and (22).

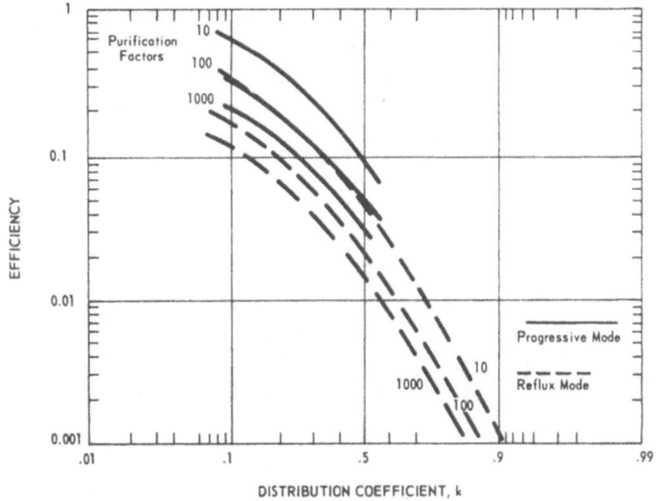

Fig. 10. Efficiency comparisons for the progressive and reflux modes for the total separation (purification plus stripping) at 90% recovery.

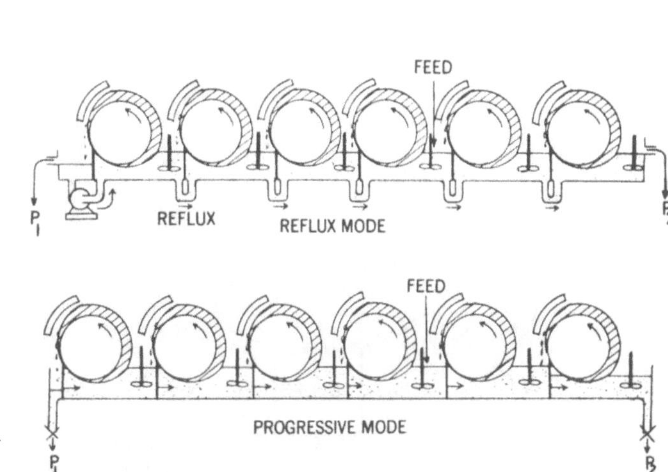

Fig. 11. Multistage rotary drum crystallization as operated in the reflux and progressive modes.

moved from the right end and restarted on the left end. At proper intervals the feed position would be transferred from the receding liquid zone to the oncoming one. Physically, the process in that form might be difficult to engineer. However, one can keep the drums stationary and thus move the charge along an infinite series of such stages. Product P_1 would be withdrawn from the left end as long as such product was of acceptable quality. Then a new stage would be added to the left end, and when its reservoir was filled, it would start producing more product. When the reservoir on the right end is reduced in quality and quantity to constitute the appropriately stripped product P_2, this would be drained and the empty stage abandoned. The feed would, of course, be put into whichever reservoir most closely approximated it in composition and would thus be intermittently moved toward the left. In this way, the entire process would progress along an infinite number of stages and hence the terminology *progressive mode*.

The logical operation of the progressive mode in this form is, of course, in a circle, Figure 12. The system contains one or two more stages than are required for the separation. When product P_2 is drained, the reservoir is cleaned and made ready to receive product P_1 when that being produced by the preceding stage is no longer acceptable. The feed point is moved when necessary, and the process as a whole progresses around the circle.

Although the progressive mode has a potential efficiency advantage of up to a factor of 2 over the reflux mode, it suffers from the disadvantage of not being steady state. The products are obtained intermittently and are of fluctuating quality, the average of which must conform to the specifications. The feed point must be changed at appropriate times. These features demand a highly programmed operating procedure along with continual monitoring of compositions. The costs of controlling a system of this type will be relatively independent of the scale of the operation, so that the larger the size, the more justifiable is the use of the progressive mode.

There are possible, of course, other multistage schemes or combinations of such schemes worthy of consideration. These will vary in their applicability to various crystallizing techniques. One example is the reflux cross-flow zone refining (or normal crystallization) process described by Pfann (6). We have not undertaken a study of the schematic optimization of that technique nor attempted a comparison with the reflux or progressive modes. Such optimization and comparison should be made, since reflux cross flow may have decided advantages in many cases.

The progressive mode is not restricted to continuously growing crystal surfaces or even to melt crystallization. The repeated normal freezing with reflux technique of Liscom, Weinberger, and Powers (7), for example, is in the progressive mode. This, though very similar topologically to the reflux cross-flow method of Pfann (6), can be shown to actually consist of two separate, though interwoven, charges of material, each of which independently undergoes its own zone fractionation (that is, schematically progressive), though with a fluctuating zone size. Multistage solvent recrystallization could be put into the progressive mode as shown in Figure 13. There is no reason, in fact, why other separation processes such as distillation or extraction could not be put into multistage progressive mode. In these cases, however, simple countercurrent contacting is far superior to any multistage processes incorporating complete phase transitions at each stage. Solid-liquid countercurrent contacting processes have been developed, for example, by Schildknecht and co-workers (8), but these have not yet enjoyed the same success as liquid-liquid and liquid-vapor contacting processes. They may ultimately prove more efficient than multistage reflux or progressive schemes, but it is unfortunately not possible to make comparisons by the theoretical or arithmetical techniques used here.

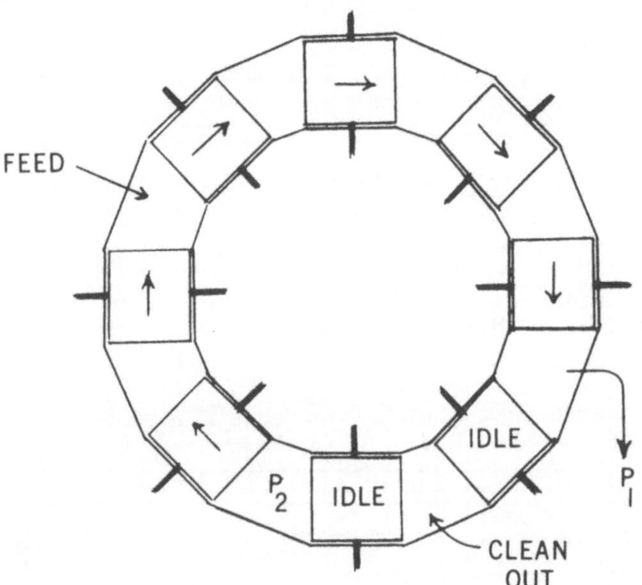

Fig. 12. Circular arrangement for rotary drum crystallizers for continuous operation in the progressive mode.

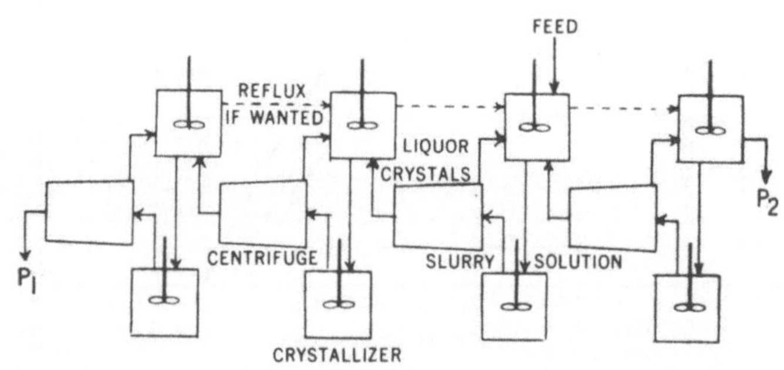

Fig. 13. Sequential arrangement of solvent recrystallization apparatus for multistage operation in either the reflux or progressive modes.

NOTATION

A = manipulative parameter defined by Equation (28)
B = manipulative variable defined by Equation (27)
C_1 = solute concentration in purified product P_1
C_2 = solute concentration in stripped product P_2
C_F = solute concentration in feed
C_L = solute concentration in liquid phase
C_S = solute concentration in solid phase
C_x = solute concentration in solid ingot at position x
E = efficiency defined in terms of feed rate and total stages
E' = efficiency defined in terms of product P_1, rate, and total stages
E_1 = efficiency of purification section alone
E_2 = efficiency of stripping section alone
F = feed, also feed rate
k = coefficient for the distribution of solute between solid and conjugate liquid
L = liquid, also total length of ingot
L_1 = length of purification section of ingot
L_1' = physical length of purification section with accumulated zones
L_2 = length of stripping section of ingot
ℓ = zone length
ℓ_S = length of solid between zones
M = number of zone lengths in an ingot
N = total number of stages
N_1 = number of purification stages
N_2 = number of stripping stages
n = any integer
P_1 = purified product
P_2 = stripped product
Q = manipulative variable defined as $(1 - T_1)/k$
R = reflux rate
S = solid, also total separation factor
S_1 = separation factor for purification section
S_2 = separation factor for stripping section
s = separation factor for a single stage
T_1 = takeoff rate of product P_1
T_2 = takeoff rate of product P_2
W = manipulative variable defined as $(1 + T_2)/k$
x = any variable; also position along ingot
Y = manipulative variable defined by Equation (39)
Z = manipulative variable defined by Equation (26)
β = manipulative variable defined by Equation (38)
θ = quantity of product P_1 per zone pass
ϕ = quantity of product P_2 per zone pass

LITERATURE CITED

1. Pfann, W. G., "Zone Melting," 2 ed., p. 1, Wiley, New York (1966).
2. *Ibid.*, Chap. 6.
3. *Ibid.*, pp. 184–190.
4. *Ibid.*, p. 187.
5. *Ibid.*, pp. 195–196.
6. Liscom, P. W., C. B. Weinberger, and J. E. Powers, *Ind. Chem. Eng.*, 73–85 (1965).
7. Schildknecht, H., *Analytica Chimica Acta*, **38**, 261–273 (1967).

ULTRAPURIFICATION OF p-TERPHENYL

G. J. Sloan
and
Vladimiro Ern

In order to provide pure material for spectroscopic studies, *p*-terphenyl was zone melted. The bottoms of ingots afforded a concentrate from which biphenyl, o-, and *m*-terphenyl and triphenylene were separated by thin layer chromatography.

Examination of delayed fluorescence along zone melted ingots indicated the presence of impurity with segregation coefficient greater than unity. Repeated zoning gave a fraction in which impurities of molecular weights 304 and 306 were detected by mass spectroscopy. These appear to be 2-phenyl triphenylene and 2,3-diphenyl biphenyl, respectively, present initially at about 5 p.p.m. This finding represents the first use of delayed fluorescence for locating impurities in a zone melted material.

Any attempt to purify a chemical carries with it the need to measure the purity attained. It is often impossible to measure purity directly, so that assay methods for specific impurities are required. In this report, we will describe the application of zone melting to the purification of *p*-terphenyl and the use of delayed fluorescence measurements to indicate the purity.

Metastable triplet excitons are a very sensitive tool to probe the purity of an organic crystal. It has been shown (*1, 2*) that triplet excitons migrate in a diffusive type of motion throughout the lattice of the crystal. For example, for anthracene the diffusion constant D was found (*2*) to be 2×10^{-4} sq. cm./sec. for the motion in the ab plane of the crystal. This gives for a typical triplet lifetime τ of 10 msec. a diffusion length $L (L \equiv \sqrt{2D\tau})$ of $20\,\mu$; that is, every exciton will scan in the average about 2×10^4 unit cells of the lattice during its lifetime.

A triplet exciton can be quenched by an impurity in the lattice as well as by another triplet exciton. This latter process of mutual annihilation of triplets leads to delayed fluorescence (*3*) from the crystal. A rough estimate of the sensitivity of triplet lifetime to impurity content can be obtained under the assumption that an impurity is as effective as a triplet in quenching triplets. That is, $\Delta\beta \approx \gamma n_I$, where $\Delta\beta$ is the change in exciton decay rate due to the presence of a concentration n_I of impurity added to the crystal, and $\gamma \approx 10^{-11}$ cc./sec. is the measured (*3*) triplet-triplet annihilation rate constant. For a change in triplet lifetime from $\tau = 10 \times 10^{-3}$ to $\tau_I = 1 \times 10^{-3}$ sec. after doping, one has

$$\Delta\beta = \frac{1}{\tau_I} - \frac{1}{\tau} = 9 \times 10^2/\text{sec.}$$

and the impurity concentration is estimated to be

$$n_I \approx \Delta\beta/\gamma \approx 10^{14}/\text{cc.}$$

that is, typically a fraction of p.p.m.

We chose to study *p*-terphenyl as part of a continuing investigation of triplet excitons in organic crystals. It should be noted, however, that this compound has been used extensively in coolants for nuclear reactors and that there has been much interest in its large scale purification.

EXPERIMENTAL PROCEDURE

ZONE MELTING

Reagent grade *p*-terphenyl (150 g.) was charged into each of two Pyrex tubes (45 cm. long by 25 mm. O.D.) in an argon atmosphere (*4*). The tubes were sealed off at a

E. I. duPont de Nemours and Company, Wilmington, Delaware.

pressure of 0.5 atm. of argon, and each was subjected to fifty zone melting passes in an automatic zone melter (5) at a travel rate of 2.5 cm./hr. The zone length was about one-twentieth of the ingot length.

Delayed fluorescence was measured at the top, middle, and bottom (T, M, and B, respectively, in Figure 1) of one of the ingots, in its sealed tube, as described below. Single crystals were grown from samples removed from the top and middle of the ingot, and again delayed fluorescence was measured. The decay of the emission from the top crystal did not conform to the pure exponential behavior expected from a single annihilative process. Consequently, zone melting was continued.

A part of segments 0005/38* was charged into a tube 40 cm. long by 13 mm. O.D. and a part of segments 0530/38 was charged into a tube 45 cm. long by 25 mm. O.D. These two tubes were subjected to fifty zone melting passes; segment 0011/33 of the first tube (TT in Figure 1) and segment 0005/37 of the second tube (TM) were combined and charged into a tube 40 cm. long by 12 mm. O.D. which was subjected to fifty zone melting passes. Again, delayed fluorescence was measured before the tubes were opened.

DELAYED FLUORESCENCE

The apparatus used to measure delayed fluorescence is shown schematically in Figure 2. Light from a xenon lamp is passed through a monochromator, filters, and collimating lenses. A rotating circular sector interrupts the light beam at fixed frequency. Any exciting light that passes through the sample is removed by filters, while fluorescence emitted at lower wavelength by the sample enters a photomultiplier. The amplified output of

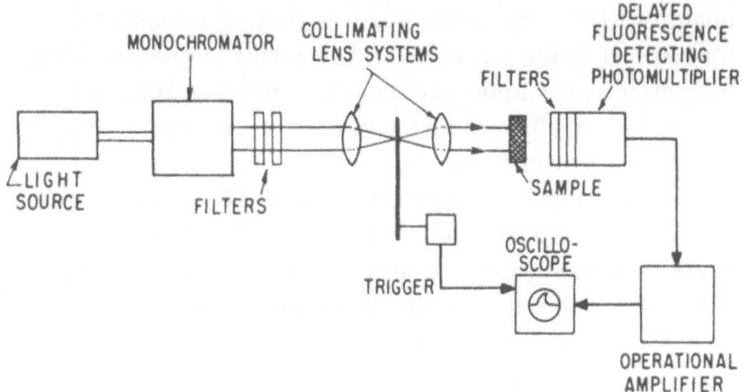

Fig. 2. Block diagram of apparatus for measurement of delayed fluorescence.

the photomultiplier is displayed on an oscilloscope, which is triggered to act synchronously with the light chopper.

This assembly provides the excitation wave form shown in Figure 3; the fluorescence resulting from the mutual annihilation of pairs of triplet excitons that are generated by the exciting light is also shown schematically in Figure 3.

The measured values of triplet lifetimes in the ingot resulting from zone melting of fractions TT plus TM are shown in Figure 4. Segment 0013/33 of this ingot showed strong, blue-white fluorescence when excited by light of 366 nm. wavelength; segment 1330/33 showed weaker, blue fluorescence, and segments 3033/33 showed no fluorescence at all.

CHROMATOGRAPHY AND MASS SPECTROSCOPY

Samples removed from several places in ingot TT plus TM were placed serially in the heated inlet of a mass

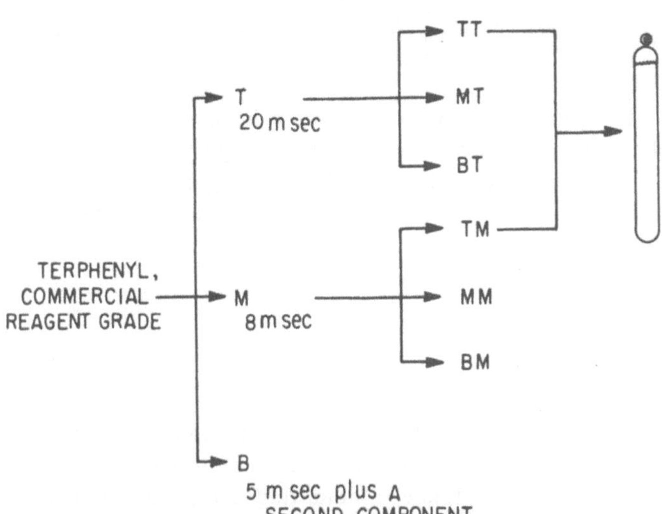

Fig. 1. Triplet lifetime τ of zone melted p-terphenyl.

*The first two digits give the distance from the top of the ingot to the top of the segment, the second two the distance to the bottom of the segment, and the digits after the slant give the length of the ingot. All distances are in centimeters.

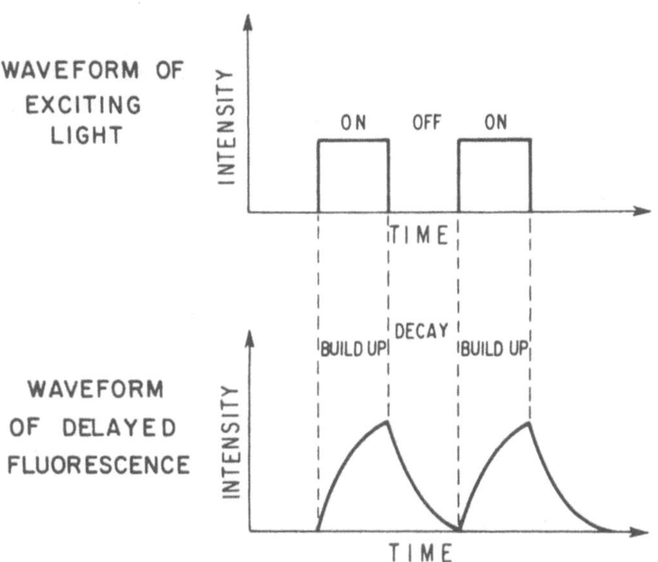

Fig. 3. Observation of delayed fluorescence.

spectrometer. The mass spectra showed no unusual features at values of m/e below that of terphenyl (230), but small peaks appeared at m/e = 304 and 306. The intensities of these peaks relative to the parent peak at 230 are shown in Figure 4.

A 43-g. sample of segment 3038/38 (B) of the original zone melting was stirred overnight with 200 ml. of acetone (reagent grade), and the slurry was filtered. The filtrate was evaporated to dryness, and a portion of the solid residue (0.8 g.) was injected into a gas chromatography under the following conditions:

column: 10% Apiezon L on Gas-Chrom Z, 2 ft. long.
carrier gas: helium, 10 cc./6 sec.
temperature: programmed 100° to 250°C. at 20°C/min.

The relative retention times of the observed peaks are tabulated below, along with the area percent of each peak as an approximate indication of the amount of each substance.

SEPARATION OF IMPURITIES FROM *p*-TERPHENYL BY GAS CHROMATOGRAPHY

Relative ret. time	Area, %	Assignment
0.45	0.9	biphenyl
0.65	2.4	*o*-terphenyl
0.82	0.6	*m*-terphenyl
1.00	95.5	*p*-terphenyl
1.15	0.5	unknown
1.41	trace	triphenylene

The impurities were identified tentatively by comparing their retention times with those of authentic samples.

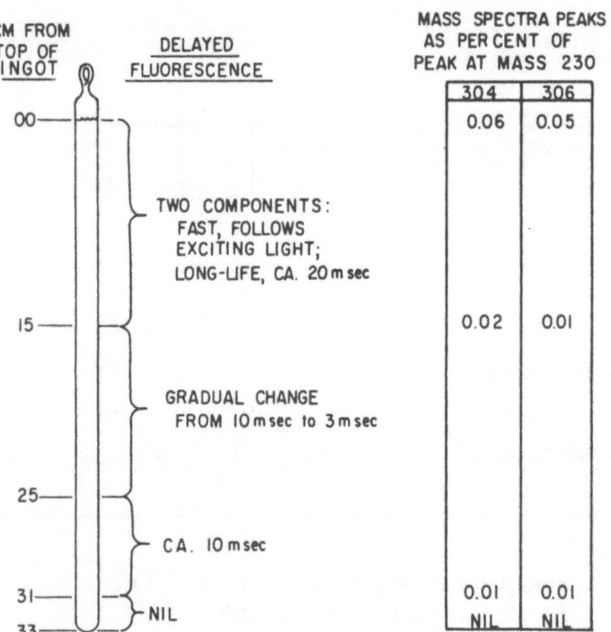

Fig. 4. Delayed fluorescence and mass spectroscopy of zone melted *p*-terphenyl (TT plum TM).

Then the individual peaks were trapped as they emerged from the chromatograph, and the traps were rinsed directly into optical cells for measurement of the absorption spectra of the solutions. The solvent was trimethylpentane (spectroscopic grade).

A sample from the top of ingot (TT plus TM) was also chromatographed under the conditions described above; only one peak appeared.

DISCUSSION OF RESULTS

Delayed fluorescence measurements gave some indication of the presence of an impurity at the top of the original zone melted ingot. The impurity in question was detectable only by mass spectroscopy after two further applications of zone melting. The maximum concentration that was measured was 0.06%. It is probable that the concentration of this substance in the original ingot was ten to one-hundred times lower, that is, 6 to 60 p.p.m.

Since the impurity has a distribution coefficient that is greater than unity (6), one may conclude that the impurity is similar in shape and size to *p*-terphenyl. That comparable amounts of impurity appear at m/e = 304 and 306 suggests that the impurity is larger than terphenyl by one phenyl group and loses hydrogen. These thoughts suggest that *o*, *p*-quaterphenyl is the impurity of m/e = 306 and that it gives 2-phenyltriphenylene photochemically in the molten *p*-terphenyl:

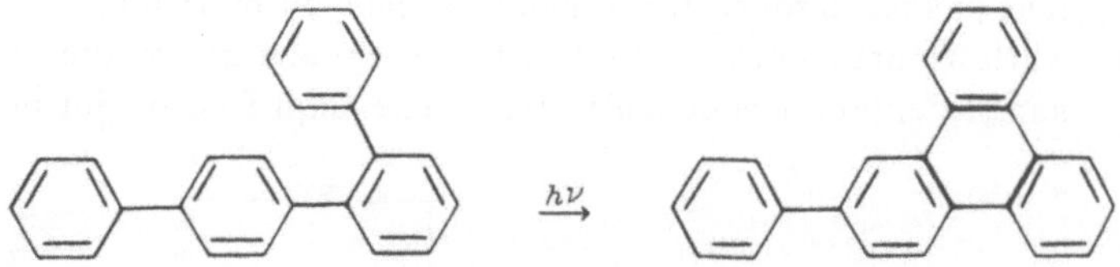

$\xrightarrow{h\nu}$

The total impurity concentration in the enriched sample that was analyzed by gas chromatography was about 4.5%. The concentration in the starting material was lower by a factor of roughly 50 and in the purest fraction by at least 1,000.

CONCLUSION

It is possible to reduce the total organic impurity content in *p*-terphenyl to less than 50 p.p.m. by zone melting. Delayed fluorescence lifetime measurements are of great assistance in detecting impurities at such low levels.

ACKNOWLEDGMENT

Contribution No. 1497, Central Research Department, Experimental Station, E. I. du Pont de Nemours and Company, Wilmington, Delaware, 19898.

LITERATURE CITED

1. Avakian, P., and R. E. Merrifield, *Phys. Rev. Letters*, **13** 541 (1964).
2. Ern, V., P. Avakian, and R. E. Merrifield, *Phys. Rev.*, **148**, 862 (1966).
3. Kepler, R. G., J. C. Caris, P. Avakian, and E. Abramson, *Phys. Rev. Letters*, **10**, 400 (1963).
4. Sloan, G. J., *Mol. Crystals*, **1**, 161 (1966).
5. ———, and N. H. McGowan, *Rev. Sci. Instr.*, **34**, 60 (1963).
6. Pfann, W. G., "Zone Melting," Chapt. 2, Wiley, New York (1966).

INFORMATION RETRIEVAL*

Size distribution analysis in continuous crystallization, Larson, M. A., and A. D. Randolph, *Chem. Eng. Progr. Symposium Ser. No.* 95, **65,** p. 1 (1969).

Key Words: Crystallization-8, Size Distribution-7, Population Balance-10, Crystal Growth-8, Nucleation-8, Continuous Crystallization-5, Fines Removal-4, Classification-4, Crystallizer Design-4, Mixed Suspension-5.

Abstract: The basic ideas of the population balance are presented along with the use of the population balance in relating system parameters and crystallization kinetics with size distribution. The principles are illustrated in the context of classification and fines destruction systems. Example problems are provided for the reader.

Mixing effects in continuous crystallization, Becker, G. W., Jr., and M. A. Larson, *Chem. Eng. Progr. Symposium Ser. No.* 95, **65,** p. 14 (1969).

Key Words: Crystallization-8, Mixing-8, Suspension-9, Size Distribution-6, 2, Crystal Size-6, Flow Model-9, Macrofluid-9, Continuous-10, Size Control-4.

Abstract: It is well known that idealized backmixed or plug flow conditions are seldom achieved in large industrial crystallizers. As a consequence, predicted size distributions are seldom achieved. This paper discusses the effect of various mixing concepts and process geometries on expected size distribution. Models assuming combinations of well-mixed and plug flow sections are discussed, as well as segregated flow models.

An experimental study of crystal size distribution in a continuous, backmixed, gas-phase reactor, Stone, Paul D., and Alan D. Randolph, *Chem. Eng. Progr. Symposium Ser. No.* 95, **65,** p. 24 (1969).

Key Words: Crystallization-8, Gas-Phase-5, Ammonium Chloride-2, Ammonia-1, Hydrogen Chloride-1, Nucleation-7, Growth Rate-7, Backmixed Suspension-10, Combustion Kinetics-4.

Abstract: An experimental study was made of the crystal size distribution produced in the gas-phase crystallization of ammonium chloride in a continuous, single-stage, backmixed reactor. Linear growth rates were investigated as a function of the holding time, and an overall crystallization mechanism for this system was deduced by means of electron microscopy data and experimental nucleation-growth rate data. Results from this work indicate that nucleation and growth rate kinetics can be studied in such gas-phase systems by measuring crystal size distributions and gas-phase compositions under a variety of operating conditions.

Continuous sugar crystallization: a chemical engineer's viewpoint, Bennett, Richard C., *Chem. Eng. Prog. Symposium Ser. No.* 95, **65,** p. 34 (1969).

Key Words: Sugar Solution-1, Sugar Crystals-2, Continuous Sugar Crystallization-4, Mechanical Stimulus-6, Nucleation Rate-7, Continuous Crystallization-8, Sugar-9, Continuous Crystallizer-10, Swenson Continuous Sugar Crystallizer-0.

Abstract: Although sugar crystallization is one of the oldest and most important crystallization activities, it is still accomplished by batch methods. A chemical engineer has much to offer in possible improvements in the current process and in the design of possible continuous processes. A workable continuous process is described, along with a discussion of the economics engendered.

Commercial urea crystallization, Bennett, Richard C., and Maarten Van Buren, *Chem. Eng. Progr. Symposium Ser. No.* 95, **65,** p. 44 (1969).

Key Words: Urea Solution-1, Bioret-1, 3, Carbon Dioxide-1, 3, Ammonia-1, 3, Crystalline Urea-2, Pure Urea Crystals-4, Water Solution-5, Crystallizer Design-6, Size of Urea Crystals-7, Crystallization-8, Changes in Nucleating Stimulus-10, Swenson Crystallizers-0.

Abstract: The crystallization of urea by a continuous method is described. The influence of biuret concentration is discussed, and various methods of correlating the size distribution data are illustrated. The results show some inconsistency by using some of the published analysis techniques. It was concluded that in this instance the process lies outside the scope of the analysis used.

Design models for continuous crystallizers with double drawoff, Hulburt, Hugh M., and Denis G. Stefango, *Chem. Eng. Progr. Symposium Ser. No.* 95, **65,** p. 50 (1969).

Key Words: A. Crystallization-8, Crystallizer-10, Continuous-0, Crystallizer Stability-8, Design-4, 8, Cooling-6, Evaporation-6, Solubility-1, 6, Nucleation-1, Growth Rate-1, Size Distribution-2, 7, Yield-7, Production Rate-2, Solution-5, Purification-4, Separation-4, Fines Control-7.

Abstract: A design model is proposed for a continuous well-stirred crystallizer in which one mother liquor stream is drawn off through a classifier to retain all but a specified fraction of the fines and a second main stream discharges mother liquor and product crystals having the average size distribution and composition of the main body. Model calculations are given to illustrate the limits of steady vs. oscillatory outputs and the stabilizing effect of selective fines removal. General procedures adaptable to computers are given for treating multiple outlets and realistic nucleation and growth kinetics in both steady and transient operating modes.

*For details on the use of these key words and the AIChE Information Retrieval Program, see *Chem. Eng. Progr.*, **60,** No. 8, p. 88 (Aug., 1964).

Dynamic behavior of the well mixed isothermal crystallizer, Sherwin, M. B., Reuel Shinnar, and Stanley Katz, **AIChE Journal, 13,** No. 6, p. 59 (November, 1967).

Key Words: A. Crystallizer-8, Crystallization-7, 8, Continuous-0, Input-6, Size-7, Particles-9, Nonlinear Equations-10, Computation-8, Time-9, Cycle-9, Amplitude-9, Fluctuations-9, Seeding-6, Stability-7, 8, Product-9, Distribution-7.

Abstract: Under some conditions, continuous crystallization exhibits cyclic changes of the particle size even with constant input conditions. A linearized stability analysis was performed to determine under what conditions this behavior can be expected. The effect of seeding on the stability limits and product distribution was also evaluated.

Growth of oriented single crystals from low melting metals, Jensen, H. J., M. J. Murtha, and George Burnet, *Chem. Eng. Progr. Symposium Ser. No. 95,* **65,** p. 95 (1969).

Key Words: Metals-1, Low Melting-1, Single Crystals-2, Oriented Metal Crystals-2, Crystallization Procedure-4, Atmospheric-5, Inert-5, Casting-Bonding-0.

Abstract: A method (casting-bonding) has been developed for the production of large single crystals of controlled shape and a lattice orientation the same as that of a small seed crystal. Molten metal is cast in a mold, at the bottom of which is located a seed crystal held under a thermal gradient so that the upper portion is molten and the lower solid. Unidirectional cooling causes crystallization to begin at the seed. A bond forms at the seed-feed metal interface, and the lattice orientation of the seed continues throughout the entire crystal. The method has been used to grow oriented single crystals of bismuth, zinc, tin, lead, cadmium, and aluminum, some as large as 3 in. in diameter by 14 in. long.

Dynamic behavior of the isothermal well-stirred crystallizer with classified outlet, Sherwin, Martin, Reuel Shinnar, and Stanley Katz, *Chem. Eng. Progr. Symposium Ser. No. 95,* **65,** p. 75 (1969).

Key Words: Simulation-8, Modeling-8, Stability-8, Dynamics-8, Cycling-8, Fluctuations-8, Crystallization-9, Crystallizer-9, 10, Isothermal-0, Classified-0, Mixed-0.

Abstract: The effect of classification on the dynamic behavior of a crystallizer is investigated by analyzing a simple model of a mixed crystallizer with classified product removal. It is shown that classification enhances the sensitivity of the crystallizer to disturbances and increases the tendency toward cyclic fluctuations. For a classified product removal model these fluctuations are observed in the production rate and the magma density.

Melt growth of refractory oxide single crystals, Dess, Howard M., *Chem. Eng. Progr. Symposium Ser. No. 95,* **65,** p. 100 (1969).

Key Words: Ruby, YAG, Czochralski, Verneuil, Melt-Growth, Single Crystal, Laser, Oxide, Congruently Melting, Crucible, Melt, Seed, Pull Rate, Rotation Rate, Thermal Gradient, Diffusion Layer, Distribution Coefficient, Dopant, Segregation.

Abstract: The use of crystals in laser technology has sparked the demand for ultra high quality oxide single crystals. Two principle melt growth techniques for producing refractory oxide single crystals are described and compared. The Verneuil process is simpler, in that no crucible is required. However, the Czochralski or melt pulling process yields higher quality crystals in larger sizes.

Crystallization of pure silicon from molten aluminum, Litz, L. M., S. A. Ring, and R. A. Mercuri, *Chem. Eng. Progr. Symposium Ser. No. 95,* **65,** p. 91 (1969).

Key Words: Silicon-1, 9, Graphite-1, 5, Silicon-Aluminum-1, 5, Thermal Convection Loop-1, 10, Silicon Pure-2, 0, Single Crystal Silicon-2, Iron-3, Calcium-3, Magnesium-3, Manganese-3, Copper-3, Boron-3, Sodium-3, Titanium-3, Chromium-3, Thermal Convection-6, 8, Temperature-6, Crystallization Rate-7, Yield-7, Purity-7, Crystallization-8, Purification-8.

Abstract: A simple thermal convection apparatus, with no moving parts, was developed to carry out the continuous recrystallization of commercial silicon from molten aluminum. At temperatures above the eutectic temperature, this process was found to produce silicon of 99.95 + % purity in a single stage.

The progressive mode of multistage crystallization, Atwood, G. R., *Chem. Eng. Progr. Symposium Ser. No. 95,* **65,** p. 112 (1969).

Key Words: A. Melt Crystallization-8, Zone Melting-8, Separation-4, 8, Purification-4, 8, Solids-1, 9, Crystal-2, Purified Product-2, Solutes-3, Impurities-3, Schematics-6, Reflux-6, Stages-6, Distribution Coefficient-6, Efficiency-7, Cost-7, Purity-7, Drum Crystallizer-10, Zone Refiner-10.

Abstract: A comparative analysis is made of two types of continuous multistage crystallization schemes. The first is a reflux process in the classical sense. The second, analogous to (and including) zone refining, is designated the progressive mode because the charge essentially progresses along an infinite series of states. The progressive mode is shown to be up to twice as efficient as the classical reflux mode.

Ultrapurification of *p*-terphenyl, Sloan, G. J., and Vladimiro Ern, *Chem. Eng. Progr. Symposium Ser. No.* 95, **65**, p. 122 (1969).

Key Words: A. Purification-8, *p*-Terphenyl-9, Gas Chromatography-10, Delayed Fluorescence-10, Melting-6, Purity-7, Excitons-9. B. *p*-Terphenyl-1, *p*-Terphenyl-2, Biphenyl-3, o-Terphenyl-3, *m*-Terphenyl-3, Triphenylene-3, 2-Phenyl Triphenylene-3, 2,3'-Diphenyl Biphenyl-3, Zone Refiner-10.

Abstract: Zone melting of *p*-terphenyl concentrated several impurities, of which biphenyl, o- and *m*-terphenyl, and triphenylene were separated by gas chromatography. Examination of delayed fluorescence in zone melted ingots revealed impurity with segregation coefficient greater than unity. Mass spectroscopy indicated that substances of molecular weights 304 and 306 were involved; these appear to be 2-phenyl triphenylene and 2,3'-diphenyl biphenyl.

MONOGRAPH SERIES

1 Reaction Kinetics by Olaf Hougen
2 Atomization and Spray Drying by W. R. Marshall, Jr.
3 The Manufacture of Nitric Acid by the Oxidation of Ammonia—The DuPont Pressure Process by Thomas H. Chilton
4 Experiences and Experiments with Process Dynamics by Joel O. Hougen
5 Present, Past, and Future Property Estimation Techniques by Robert C. Reid

SYMPOSIUM SERIES

AEROSPACE

3 Rocket and Missile Technology
) Applications of Plastic Materials in Aerospace
2 Chemical Engineering Techniques in Aerospace
61 Aerospace Chemical Engineering
63 Aerospace Life Support

BIOENGINEERING

6 Chemical Engineering in Medicine
69 Bioengineering and Food Processing
84 The Artificial Kidney
86 Bioengineering . . . Food
93 Engineering of Unconventional Protein Production

FLUIDIZATION

8 Fluidization
62 Fluid Particle Technology
67 Fluidized Bed Technology

ENERGY CONVERSION AND TRANSPORT

5 Heat Transfer, Atlantic City
) Heat Transfer, Research Papers for 1954
7 Heat Transfer, St. Louis
3 Heat Transfer, Louisville
) Heat Transfer, Chicago
30 Heat Transfer, Storrs
32 Heat Transfer, Buffalo
41 Heat Transfer, Houston
57 Heat Transfer, Boston
59 Heat Transfer, Cleveland
64 Heat Transfer, Los Angeles
75 Energy Conversion Systems
79 Heat Transfer with Phase Change
82 Heat Transfer—Seattle
87 Advances in Cryogenic Heat Transfer
92 Heat Transfer—Philadelphia

ION EXCHANGE

4 Ion Exchange
24 Adsorption, Dialysis, and Ion Exchange

KINETICS

4 Reaction Kinetics and Transfer Process
5 Reaction Kinetics and Unit Operations
72 Recent Advances in Kinetics
73 Kinetics and Catalysis
89 Recent Advances in Kinetics and Catalysis

MATHEMATICS

1 Advances in Computational and Mathematical Techniques in Chemical Engineering
7 Applied Mathematics in Chemical Engineering
42 Statistical and Numerical Methods in Chemical Engineering
50 Optimization Techniques

NUCLEAR ENGINEERING

1 Part I (Ann Arbor)
2 Part II (Ann Arbor)
3 Part III (Ann Arbor)
19 Part IV
22 Part V
23 Part VI
27 Part VII
28 Part VIII
47 Part X
51 Part XI
53 Part XIII
56 Part XIV
60 Part XII
65 Part XV
68 Part XVI
71 Part XVII
80 Part XVIII
83 Part XIX
94 Part XX

PHASE EQUILIBRIA

2 Pittsburgh and Houston
3 Minneapolis and Columbus
6 Collected Research Papers
81 Phase Equilibria and Related Properties
88 Phase Equilibria and Gas Mixtures Properties

POLLUTION CONTROL

5 Pollution and Environmental Health
45 Pollution Control Engineering

PROCESS DYNAMICS

6 Process Dynamics and Control
6 Process Systems Engineering
55 Process Control and Applied Mathematics

PROCESSING TECHNOLOGY

5 Mineral Engineering Techniques
0 Liquid Metals Technology—Part I
4 Petrochemicals and Petroleum Refining
43 Recent Advances in Ferrous Metallurgy
49 Polymer Processing
54 Hydrocarbons from Oil Shale, Oil Sands, and Coal
85 Fossil Hydrocarbon and Mineral Processing

THERMODYNAMICS

7 Applied Thermodynamics
44 Thermodynamics

TRANSPORT PROPERTIES

6 Mass Transfer
8 Selected Topics in Transport Phenomena
77 Fundamental Research on Heat and Mass Transfer

WATER

8 Water Reuse
90 Water—1968

MISCELLANEOUS

1 Ultrasonics—Two Symposia
6 Chemical Engineering Education—Academic and Industrial
8 Chemical Engineering Reviews
0 Small-Scale Equipment for Chemical Engineering Laboratories
74 Physical Adsorption Processes and Principles
76 High Pressure Technology
91 Unusual Methods of Separation
95 Crystallization from solutions and melts